INTRODUCING

# The Universe

Felix Pirani and Christine Roche

Edited by Richard Appignanesi



ICON BOOKS UK  TOTEM BOOKS USA

This edition published in the UK
in 1999 by Icon Books Ltd.,
Grange Road, Duxford,
Cambridge CB2 4QF
email: icon@mistral.co.uk
www.iconbooks.co.uk

Distributed in the UK, Europe,
Canada, South Africa and Asia by the
Penguin Group: Penguin Books Ltd.,
27 Wrights Lane, London W8 5TZ

This edition published in Australia
in 1999 by Allen & Unwin Pty. Ltd.,
PO Box 8500, 9 Atchison Street,
St. Leonards NSW 2065

Previously published in the UK and
Australia in 1993 under the title
*The Universe for Beginners*

Reprinted 1996, 1997

First published in the United States
in 1994 by Totem Books
Inquiries to: PO Box 223,
Canal Street Station,
New York, NY 10013

In the United States,
distributed to the trade by
National Book Network Inc.,
4720 Boston Way, Lanham,
Maryland 20706



Originating editor: Richard Appignanesi



Printed and bound in Australia
by McPherson's Printing Group, Victoria

# THE UNIVERSE FOR BEGINNERS

The Universe is everything that exists.

Cosmology is the scientific study of the Universe.

More than can be seen. Beyond the Earth's atmosphere, only the Sun, the Moon, a few thousand stars, half a dozen planets, a few nebulae*, sometimes a comet, can be seen with the naked eye. Astronomers augment the naked senses with telescopes, spectrographs, radio dishes, tanks of cleaning fluid. They place instruments on rockets and satellites, and deep underground. Their 'seeing' is not the same as everyday seeing.

*There is a Glossary of unfamiliar and technical words, beginning on page 170

Ground rules for this book: The Universe is as it is because it was as it was, not because Anybody made it so.

As far as this book is concerned, the Universe is a physical system, not a creation of the gods. It was not designed by a Great Designer. The sciences for studying it are physical sciences: astronomy and physics. It makes sense to ask

A star does not have intent. A galaxy does not act in order to bring about any result.

The Universe has a history. Stars form, evolve, radiate energy and change size. Galaxies spread apart: the Universe expands. You do not see stars or galaxies as they are now, but as they were when they emitted the light which is seen now.

outward in space =

# Looking Backward in Time

Light from the Andromeda galaxy, for example, takes about 2,300,000 years to reach the Earth, so that Andromeda is seen, here and now, as it was about 2,300,000 years ago. The distance light travels in a year is called a light-year (ly); Andromeda is about 2,300,000 light-years away.

Which may seem obvious enough. Four hundred years ago few people believed it.

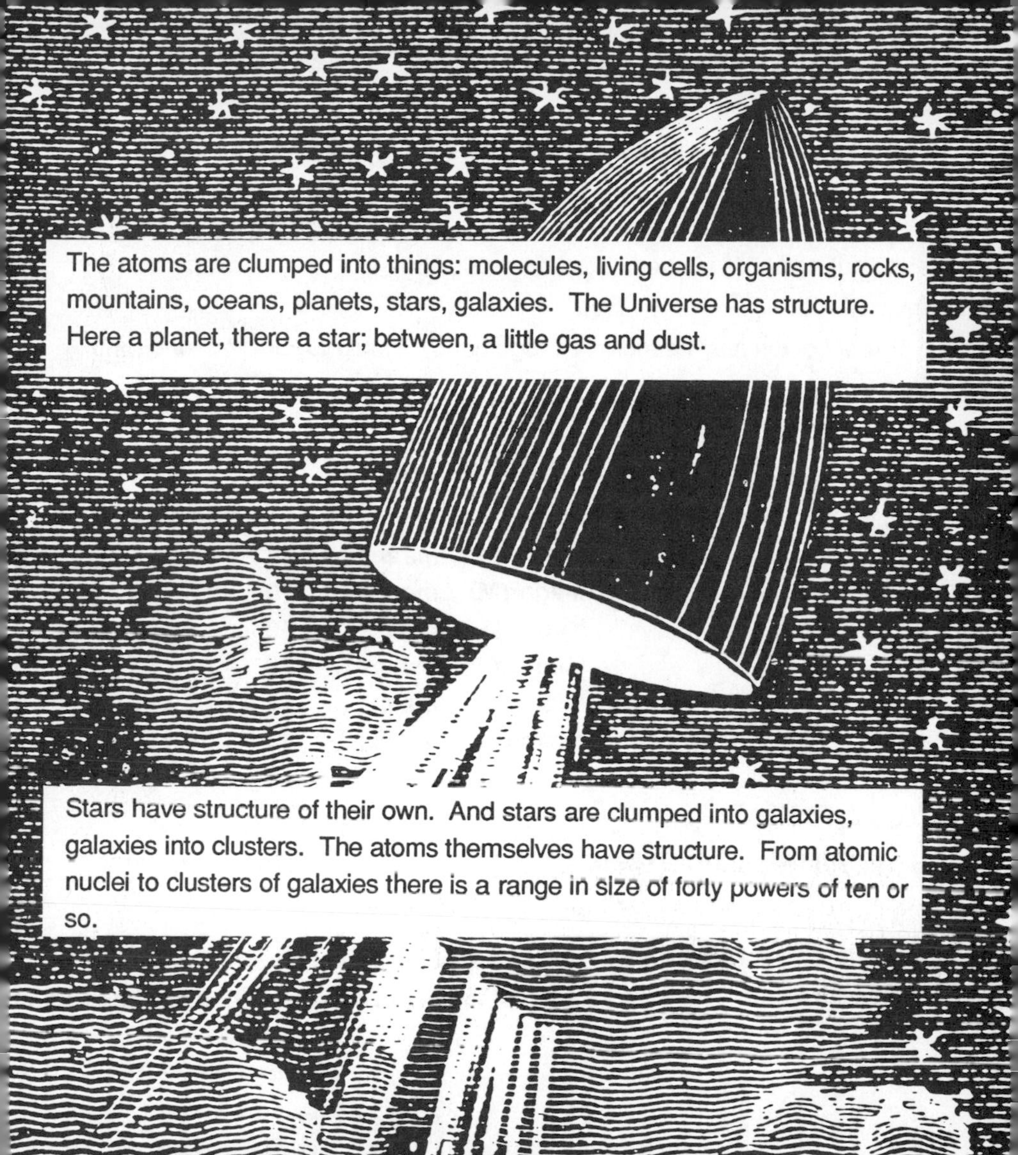

The atoms are clumped into things: molecules, living cells, organisms, rocks, mountains, oceans, planets, stars, galaxies. The Universe has structure. Here a planet, there a star; between, a little gas and dust.

Stars have structure of their own. And stars are clumped into galaxies, galaxies into clusters. The atoms themselves have structure. From atomic nuclei to clusters of galaxies there is a range in size of forty powers of ten or so.

It is the structure which makes the Universe interesting. Who would care about a Universe consisting of nothing but thin soup? Nobody, for in such a Universe there would be nobody to care about anything.

# Putting the numbers in.

The exponent notation is very useful for writing powers of ten: If n is any whole number then $10^n$ means 1 followed by *n* zeroes and $10^{-n}$ means 1 divided by $10^n$. For example $10^9$ is a billion (1,000,000,000, a thousand million) and $10^{-6}$ is one millionth (1/1,000,000).

Powers of ten may be multiplied together by adding exponents: for example

$$10^9 \times 10^{-6} = 10^3.$$

Some particular powers of ten have special names:

| | | |
|---|---|---|
| **Giga (G)** | *means* | $\mathbf{10^{9}}$ |
| **Mega (M)** | *means* | $\mathbf{10^{6}}$ |
| **Kilo (k)** | *means* | $\mathbf{10^{3}}$ |
| **Milli (m)** | *means* | $\mathbf{10^{-3}}$ |
| **Micro (μ)** | *means* | $\mathbf{10^{-6}}$ |
| **Nano (n)** | *means* | $\mathbf{10^{-9}}$ |

So that, for example, 3 Gly is $3 \times 10^9$ light-years, and 5 μm is $5 \times 10^{-6}$ metres.

Later, $10^{-43}$ seconds comes in. $10^{-43}$ seconds is a tenth of a millionth of a millionth of a millionth of a millionth of a millionth of a millionth of a millionth of a second.

# The scale of the Universe

Imagine the scale of the Universe reduced by a factor $10^{10}$.

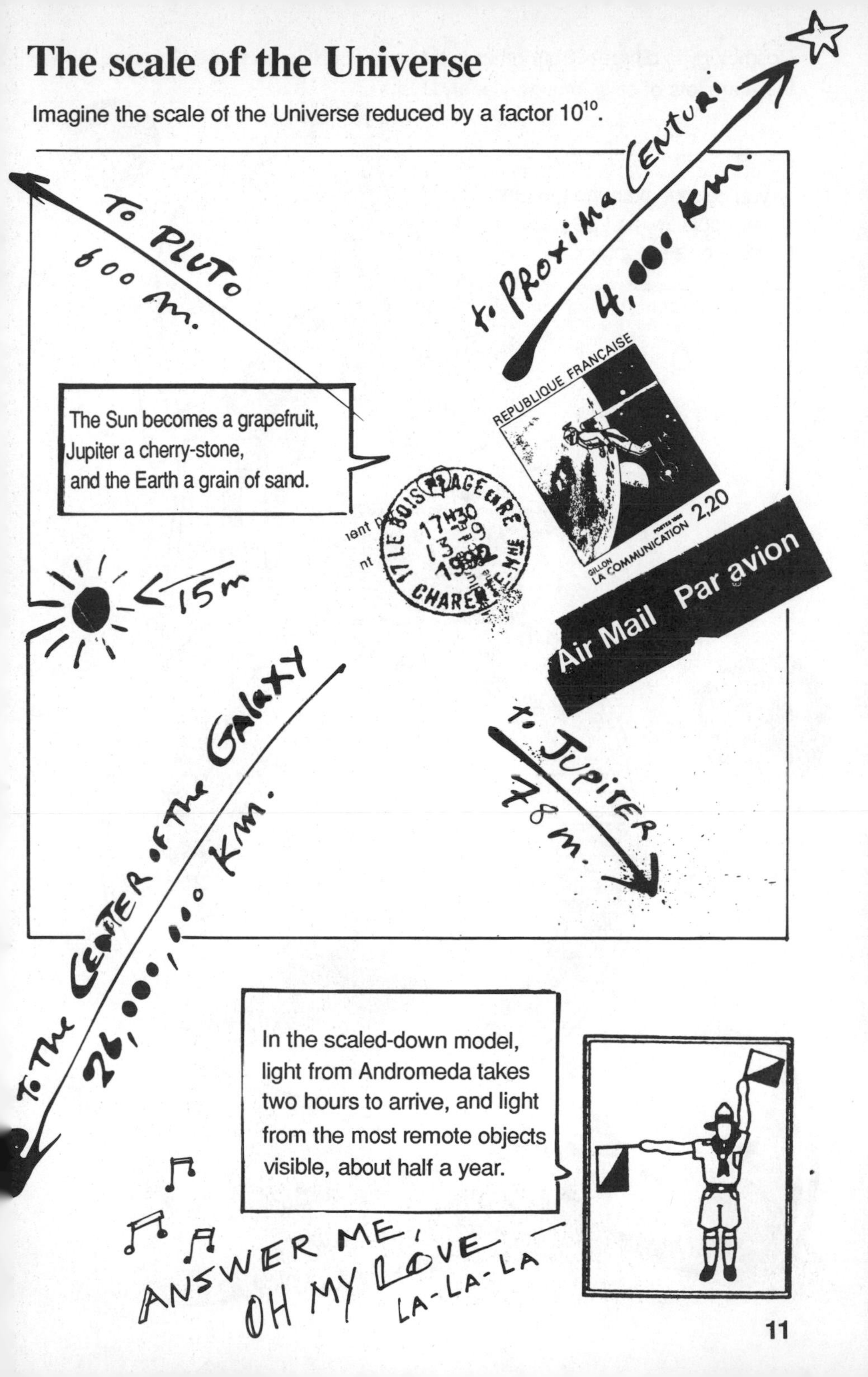

Cosmology is different from other sciences. There is only one Universe, but there are lots of cells, mountains, and stars.

— you cannot examine the Universe from outside, as you can examine a cell or a mountain.

little chance of experiment from here.

I ASK YOU TOM—
HOW DO YOU KNOW
WHAT STARS ARE MADE OF—
HOW DO YOU KNOW
HOW BIG THE GALAXIES ARE—
WHAT DO YOU MEAN, THE UNIVERSE EXPANDS?
Virginia Woolf
THE ANSWERS CHANGE, MY DEAR, COSMOLOGY TOO HAS A HISTORY
T.S.Eliot
Do I dare
Disturb the Universe?
In a minute there is time
For decisions and revisions which a minute will reverse.

# Egyptian Cosmology

The sun-god Ra, having created himself, was united with his own shadow and begat twins, Shu, god of the air, and Tefnut, goddess of rain. Shu and Tefnut, united, also produced twins, the earth-god Geb and the sky-goddess Nut. Geb and Nut in turn were united, but this angered their jealou

grandfather, Ra. Ra ordered Shu to separate them, and to hold Nut high above the Earth, as befitted a sky-goddess. Since then, Nut touches the Earth only with the tips of her fingers and toes. Her belly, covered with stars, which are her children, forms the arch of heaven.

# The Greek connection

Western cosmology begins with the philosophers of Miletus, an ancient Greek colony in Asia Minor. Thales, traditionally the earliest of them, was reputedly the first to abandon mythical versions of cosmology and work out a secular picture of the Universe.

Thales may have been a practical man, too.

Anaxagoras of Clazomenae was first in the West to state clearly that

For this he was prosecuted for impiety. The Athenian leader Pericles made a speech in his defence.

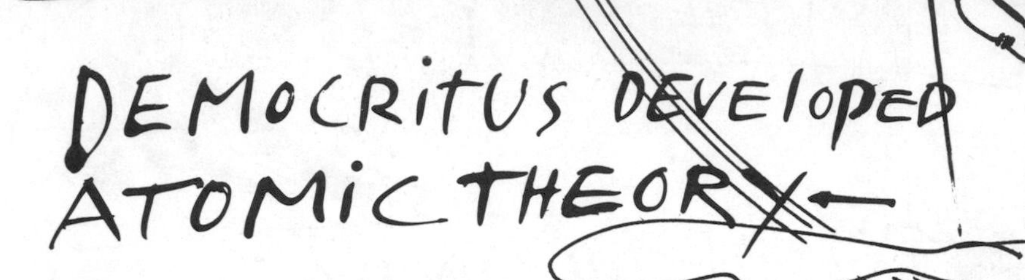


**My atoms are indivisible, and of all sorts of shapes and sizes, but so small as to be invisible.**

For Plato the stars were living beings, divine, eternal. He maintained that the actual motions of the heavenly bodies must be resolved into uniform circular motions. This dictum prevailed in astronomy until the end of the sixteenth century.

Plato: GOD IS ALWAYS BUSY WITH GEOMETRY—

Aristotle taught that the Universe has always existed. According to Aristotle, the Earth is spherical, and at rest at the centre of the Universe. The Universe is large but finite, and bounded by the sphere of the fixed stars, the primum mobile, with nothing outside. Within the primum mobile are nested spherical shells, all centred on the centre of the Earth, and all consisting of aether, the heavenly element, which is a crystalline solid.

Aether is an invisible, weightless substance, purer than fire. The planets and stars themselves are made from it. The planets, the Sun and the Moon are a embedded in aetherial shells. The shell containing the Moon is the lower boundary of the celestial region.

The motion of the inner shells is driven through mechanical linkages by further shells from the primum mobile, which is the source of all motion in the Universe. Altogether 55 spheres are required for the complete mechanism.
BUSY WITH GEOMETRY OH MASTER, OH SUN OF MY GALAXY?

For Aristotle, each element has its natural position. The natural motion of the Earth as a whole, like that of each of its parts, is towards the centre of the Universe; that is the reason why it is now lying at the centre.

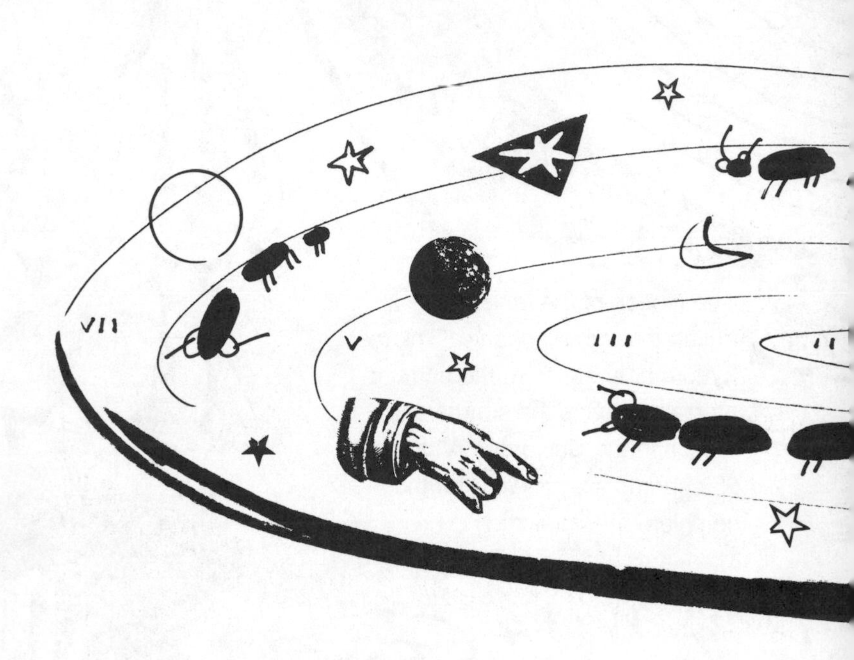


This notion of a fixed central Earth is inseparable from Aristotle's theory of mechanics. A moving Earth would also break down the absolute distinction which he drew between celestial and terrestrial matter. Worse still, if the Earth is merely a planet, the prospect of a plurality of worlds, with other inhabited planets, is opened up.

For the practical navigator, the stars hardly move relative to one another in the sky. But the stellar sphere moves daily around the Earth, and the Sun, the Moon and the five planets known in ancient times, Mercury, Venus, Mars, Jupiter and Saturn, move relative to the stars.

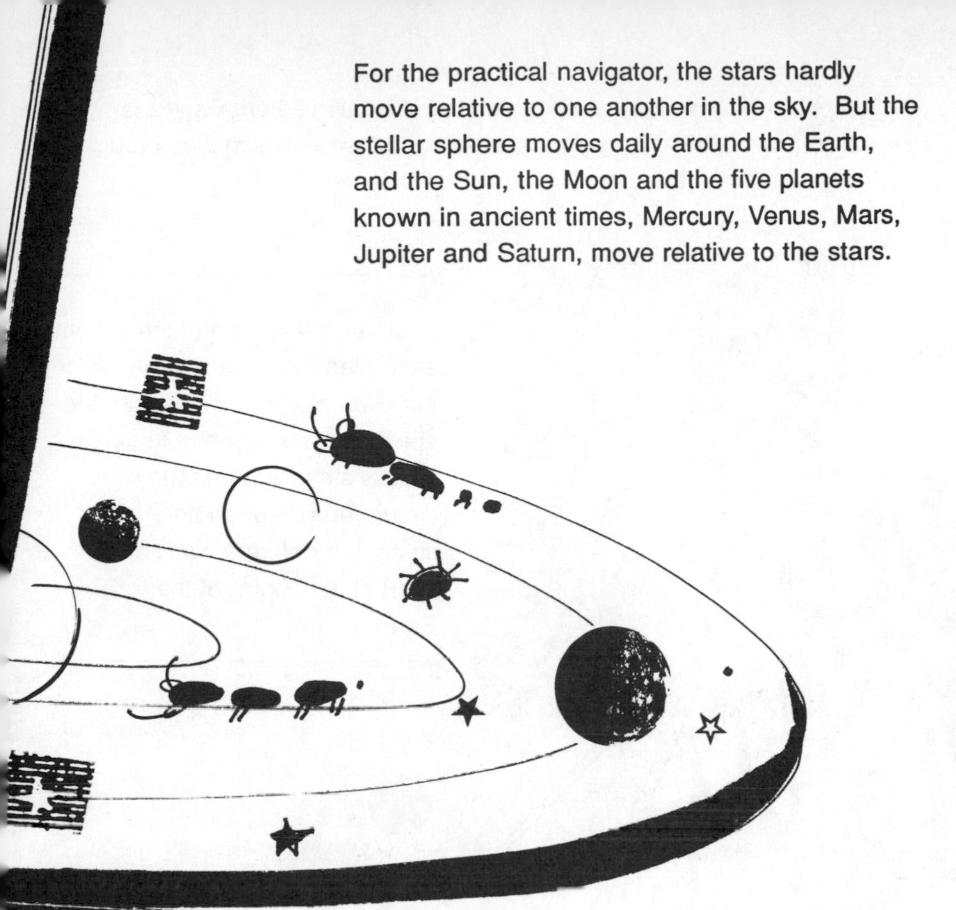

Vitruvius: Imagine seven ants crawling around circular channels on a potter's wheel. The wheel turns in the opposite direction. The ant which is nearest to the centre will finish its journey soonest, while the ant which goes round the outer edge of the wheel, with a longer circuit, will be much slower in completing the course, even if it move just as quickly as the others. In the same way the planets struggle on against the course of the firmament.

To describe the planetary motions, first Apollonius of Perga and then Hipparchus developed a new mathematical scheme, which was brought to its highest state by Ptolemy.

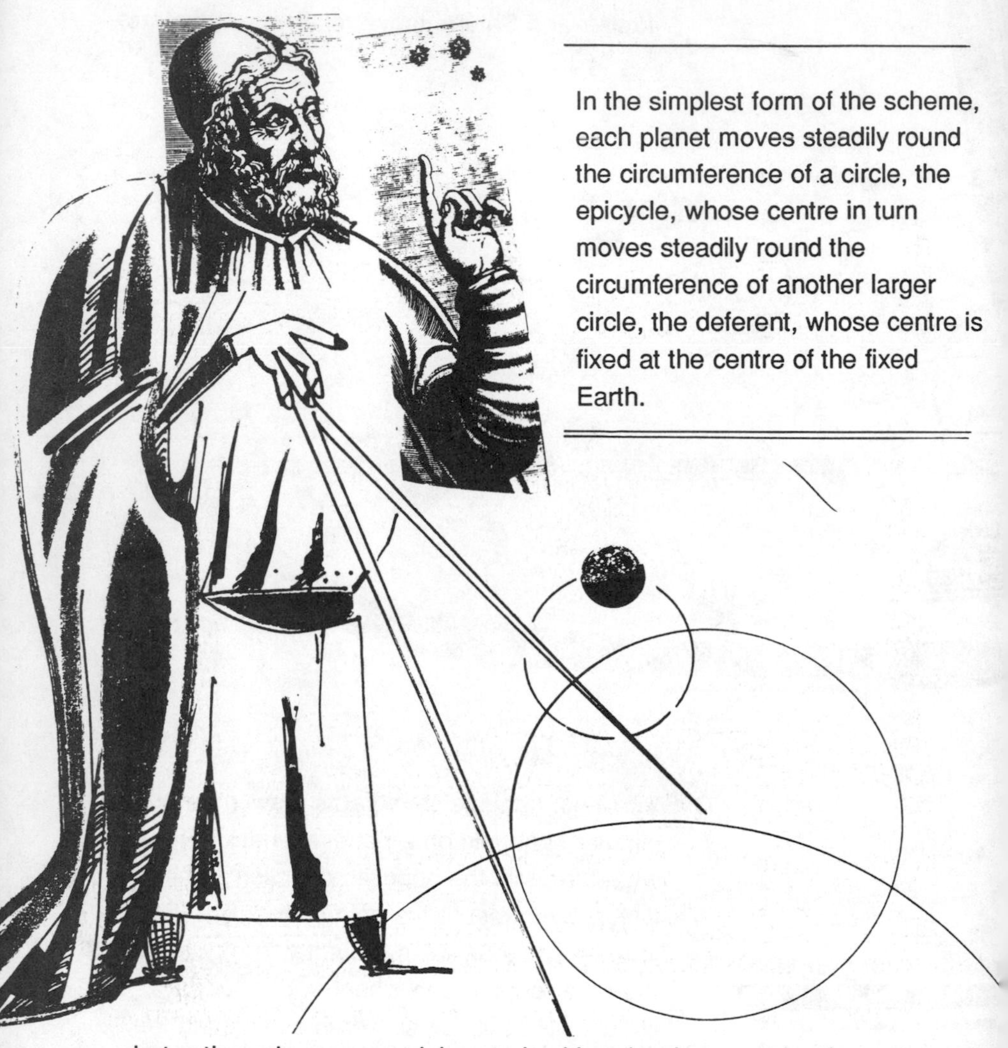

In the simplest form of the scheme, each planet moves steadily round the circumference of a circle, the epicycle, whose centre in turn moves steadily round the circumference of another larger circle, the deferent, whose centre is fixed at the centre of the fixed Earth.

Later the scheme was elaborated, with epicycles on epicycles on deferents. It lasted until the time of Copernicus — about 1400 years.

# The Biblical Universe

The Christian doctrines, and even the Jewish doctrines which preceded them, have always been statements about spiritual reality, not specimens of primitive physical science. *C.S.Lewis*

The Reverend Harry Emerson Fosdick: In the Scriptures the flat earth is founded on an underlying sea; it is stationary; the heavens are like an upturned bowl or canopy above it; the circumference of this vault rests on pillars; the sun, moon and stars move within this firmament. There is a sea above the sky, and through the "windows of heaven" the rain comes down. This is the world view of the Bible.

The Chinese scholar Ko Hung was one of those who recognized the possibility of an infinite, largely empty Universe.

The Jains, adherents of a South Asian religion, have numbers of gods, in various heavens, but they do not believe that any of these gods created the Universe. They think that the Universe has always existed and could not be destroyed.

---

*Some foolish men declare that a Creator made the World. This is not a good idea. Have nothing to do with it.*

*If a god made the World, where was he before he made it? If he needed nothing to stand on, where is he now?*

*If no one was needed to make the stuff that the World was made from, then the World could have come about by itself, without the need for anyone to make it.*

*If he created the World out of love for living things and need for them, why did he not make the World completely happy and free of misfortune?*

*The idea that the World was created by a god makes no sense at all.*

He was still being condemned at the beginning of the thirteenth century, but within a few years the Dominican friar Thomas Aquinas and others had embarked on a reconciliation.

Genesis says that God created the Universe. How can Aristotle say that it always existed?

The Bible uses metaphors which are addressed to the ignorant.

Aristotle's cosmology became dominant. Dante Alighieri celebrated it in his epic poem *The Divine Comedy*.

* WATCH DOGS OF GOD: A PUN ON DOMINICANS.

# Copernicus in the year of his death

*De Revolutionibus Orbium Cœlestium* was published with a low-profile preface by the Lutheran theologian Andreas Osiander.

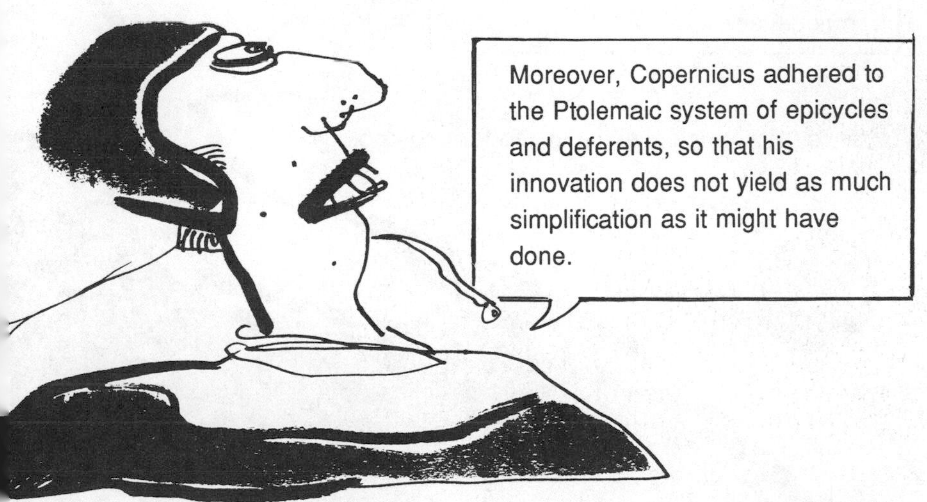

Copernicus's important contribution was to displace the Earth from the centre. The idea that the Earth is not in a specially favoured position is sometimes called the Copernican principle. It leads naturally to the idea that the Earth is in a typical position in the Universe, which is the cosmological principle.

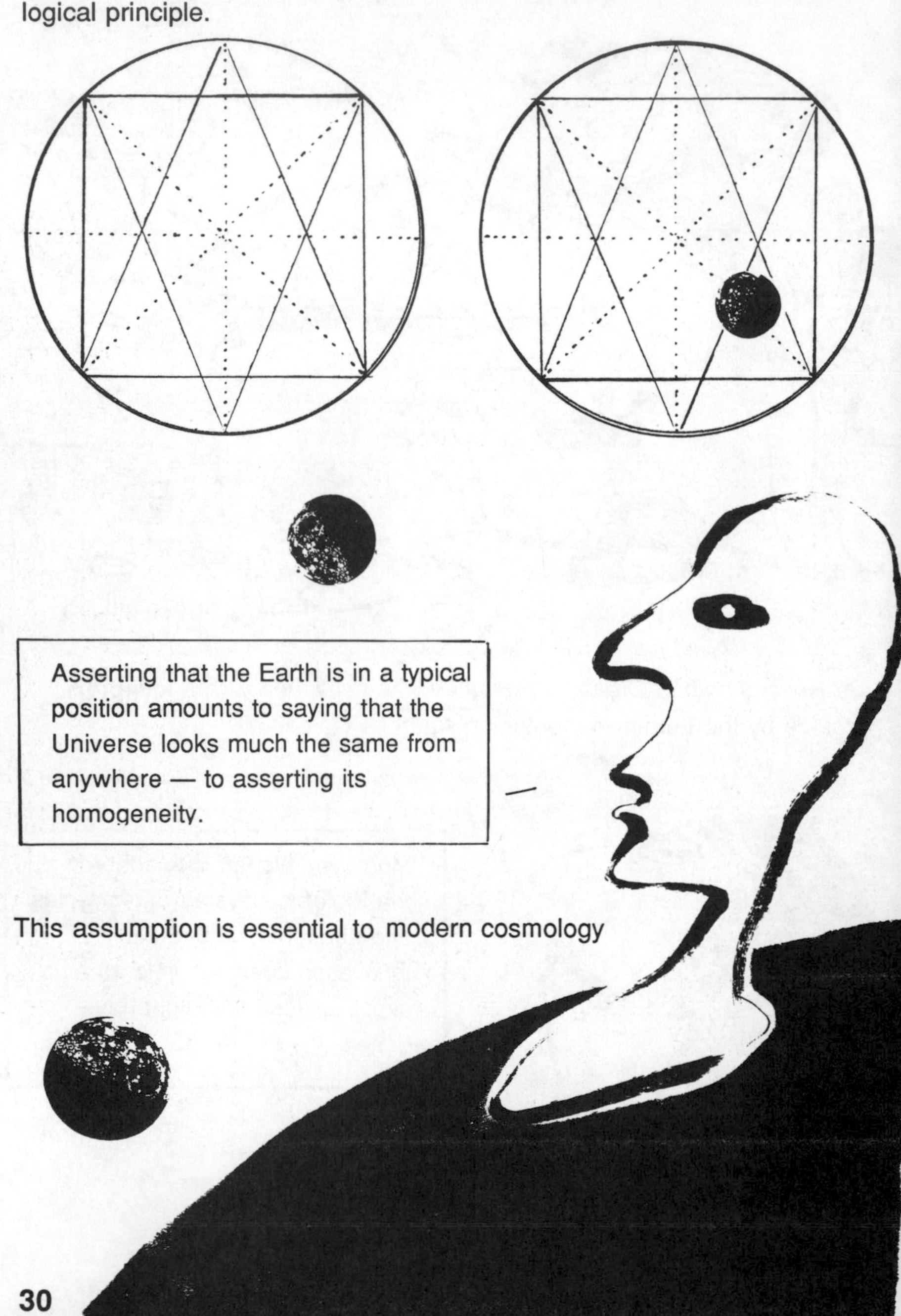


This assumption is essential to modern cosmology

Who will venture to place the authority of Copernicus above that of the Holy Spirit?

This fool wishes to reverse the whole science of astronomy; but sacred scripture tells us that Joshua commanded the Sun to stand still, and not the Earth.

...BUT HAVE YOU HEARD OF COPERNICUS!

In 1572 the astronomer Tycho Brahe discovered a new star, or ...

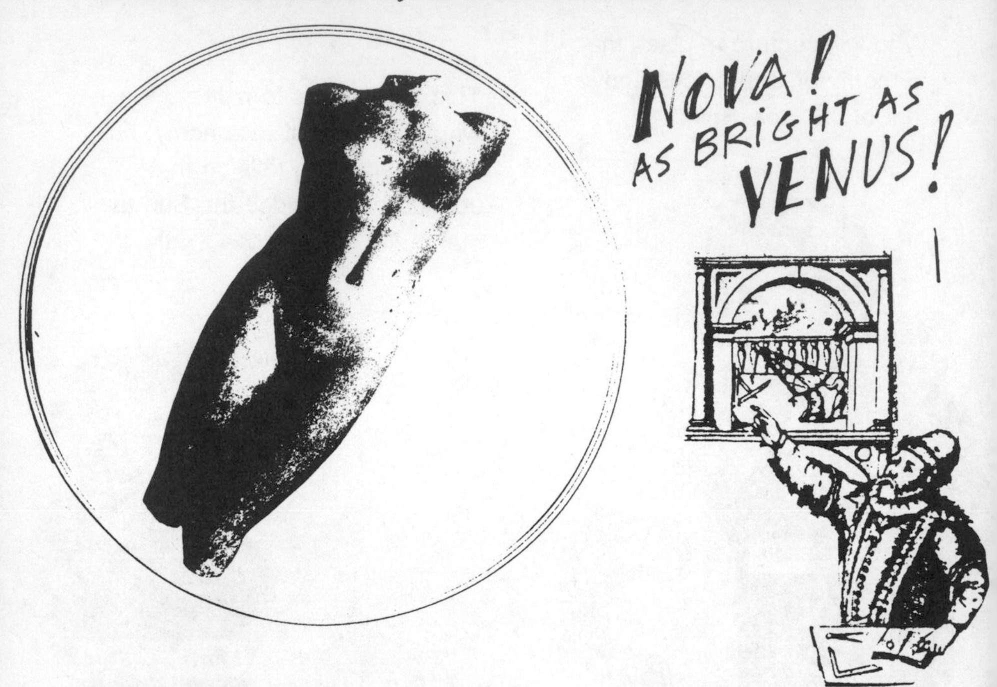


Novae are stars which become visible, or more noticeable, to observers on Earth as the result of a sudden increase in brightness. This one was probably what is now called a Type I supernova.

This nova must lie beyond the planets. And some comets are further away than the Moon. These discoveries put in question the unalterability of the heavens. I shall abandon Aristotle's crystalline spheres. But I don't need the motion of the Earth for my version of the Solar System, whatever Copernicus says.

THEY SAY HE LOST HIS NOSE IN A DUEL.... SOME STAR!

The theory of planetary motion was enormously simplified by Johannes Kepler, who was assistant to Tycho and his successor as imperial mathematician to Rudolf II, the Holy Roman Emperor. Kepler worked on a mass of planetary data accumulated by Tycho. At first he followed the tradition that all planetary motions must be built of circles, but after laborious calculations he concluded that

**This rejection of Plato's dictum was a major step towards modern astronomy.**

**Tycho, like most astronomers before him, seems to have believed in astrology. Kepler also made some money as an astrologer. He cast at least 800 horoscopes.**

# Kepler's laws

**1.** Each planet moves in an ellipse, with the Sun at one focus.

**2.** The radius vector from the Sun to the planet sweeps out equal areas in equal times.

**3.** The ratio

$$\frac{\textit{cube of the semi-major axis of the ellipse}}{\textit{square of orbital period of the planet}}$$

is the same for all the planets.

WHAT'S AN ELLIPSE?

An oval. You can draw one like this, with two drawing pins and some string. Each of the points where the pins are is called a *focus* (plural: foci).

WHAT'S THE RADIUS VECTOR?

The line from the focus (where the Sun is) to the planet. It traces out the shaded area. When the planet is nearer to the Sun, it has to go further to trace out the same area in a certain time, so it goes faster. When it's further way, it goes slower.

SEMI-MAJOR AXIS

WHAT'S THE ORBITAL PERIOD?

The time it takes the planet to go all the way around once. The further away from the Sun a planet is, the longer it takes to go round. The Earth takes a year. Jupiter, which is about 5.2 times as far from the Sun, takes about 11.86 years (and $5.2^3$ is about equal to $11.86^2$, in agreement with Kepler's third law).

Galileo Galilei, working in Padua, first turned a telescope on the stars in 1609. He quickly discovered the satellites of Jupiter, the mountains of the Moon, sunspots, and the phases of Venus. The last of these discoveries was overwhelming evidence that Venus's orbit was centred on the Sun, but did not by itself prove anything about the Earth's orbit.

# Domini Canes strike again...

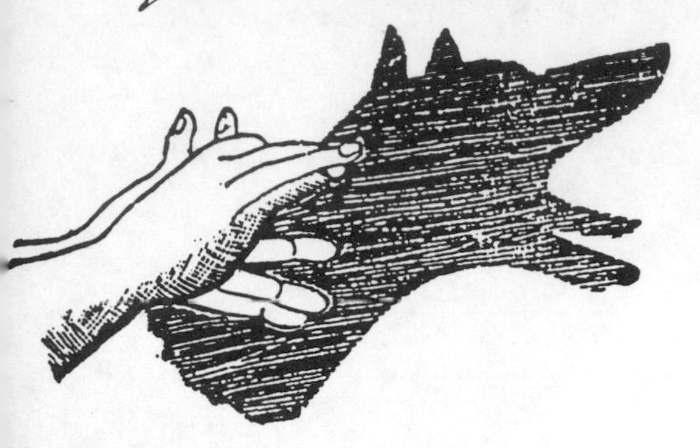

We forbid the teaching of Galileo's discovery of sunspots at the University of Louvain.
And Innsbruck.
And Pisa.
And Salamanca.

The Aristotelian distinction between terrestrial and celestial had been obliterated. The perfection of the Sun had been put in question.

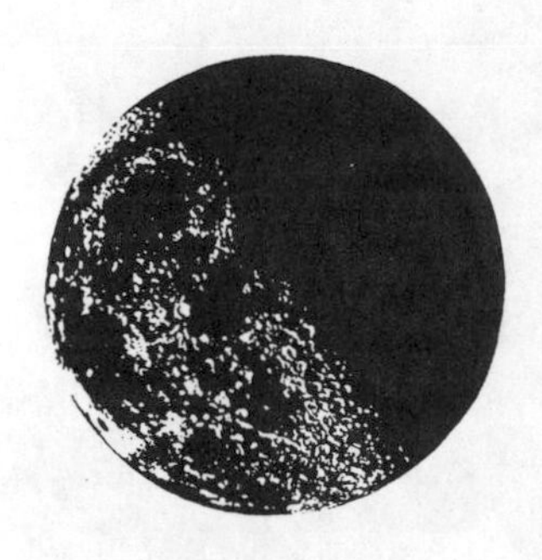

*Father Clavius*

Perhaps I can reconcile the new telescopic observations with the Aristotelian theory of the perfect sphericity of the moon. Maybe the valleys and mountains of the moon are covered by a crystalline substance, absolutely transparent, which is distributed so as to give the Moon a perfectly smooth surface.

Galileo was determined to persuade the Church authorities that they must come to terms with the new Copernican astronomy. But some Dominicans were out to get him. In 1616 the experts of the Inquisition were required to consider two propositions:

> **1** that the Sun is the centre of the Universe and hence is immovable
>
> **2** that the Earth is not the centre of the Universe, nor motionless, but moves, and turns daily on its axis.

---

## The Experts:

The first proposition is foolish and philosophically absurd, and formally heretical inasmuch as it expressly contradicts many passages in Holy Scripture.

The second proposition receives the same censure in philosophy and is declared at least erroneous in faith.

The censors suspended the books of Copernicus, until "corrected". Books that taught the same doctrine were prohibited. Galileo was at that time protected by the powerful and Catholic Medici family, and his works were not specifically listed in the condemnation.

*Cardinal Bellarmino* :

I warn him that the doctrine attributed to Copernicus, that the Earth moves round the Sun and that the Sun is fixed in the centre of the universe without moving from west to east, must not be defended or held.

In 1632 he published, with the Church's imprimatur, his *Dialogue concerning the two chief world systems*. At the time, the Pope, Urban VIII Barberini, previously a Galileo supporter, was in the midst of coping with a complex political crisis.

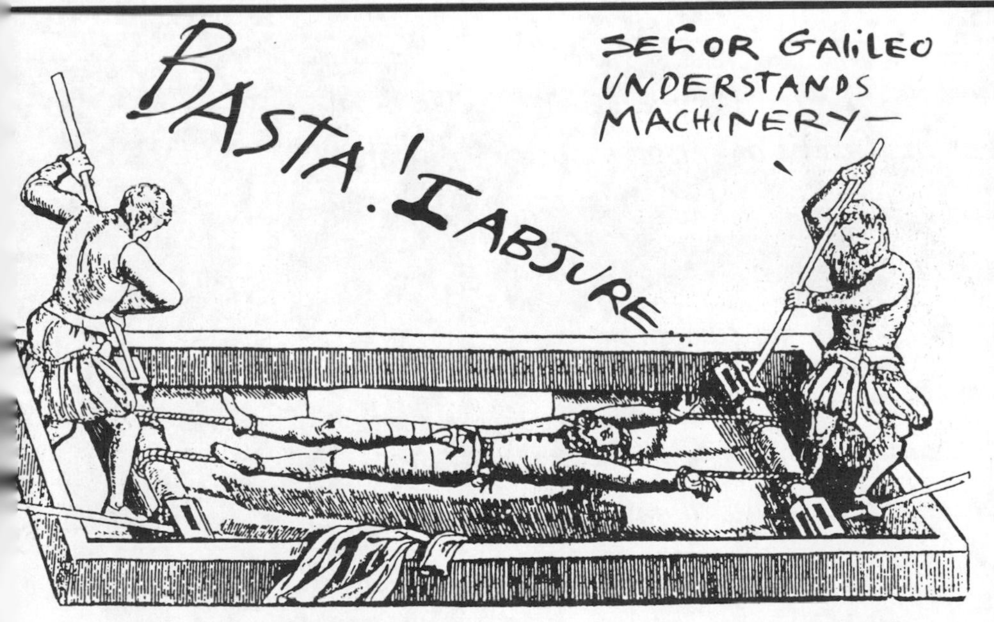

and Reverend Lord Cardin... ...tor

...mmonwealth, having before my e...

...always believed, do now believe, and ...

...ched by the Holy Catholic and Aposto...

...as lawfully given to me by this Holy Offic...

...ntre of the World and immovable, and ...

...st not hold, defend, or teach the said f...

...een notified that the said teaching is c...

...h the ... condemned d...

... solut... these, ...

...d and ... ed th...

...ot the centre and m... Therefore, ...

...ristian this vehem... suspicion justly conce...

... curse, and dete... th... aforesaid error...

... contrary to ... Church; a...

... or in writ... that might ...

...ow any heretic or person suspe...

...of ... ce wh... hall ...

...ntity ...

...ve... my said prom...

...ments ... gated by ...

...h delinquents. ...

...alilei Galilei, have abjured, sworn, prom...

...d the present document of my abjuration ... nd

... Sopra Minerva in Rome, this ... ine,

# Galileo wins, 359 years late

AFTER 359 years, the Catholic Church has admitted that it was wrong to condemn Galileo (left) for asserting that the earth orbits the sun.

At the weekend the Pope accepted the results of a commission of the pontifical academy of sciences he set up 13 years ago to study the case.

The Inquisition condemned Galileo in 1633 for backing a theory of the astronomer Copernicus because it clashed with the Bible.

"Better late than never," said Margherita Hack, director of astronomy of the University of Trieste. — Reuter.

The development of astronomical instruments and astronomical theory proceeded hand-in-hand during the eighteenth century. With the Industrial Revolution came improved methods of glass-making and metal-working, both important for astronomy. Towards the end of the century, William Herschel constructed larger and more powerful telescopes than had been possible before.

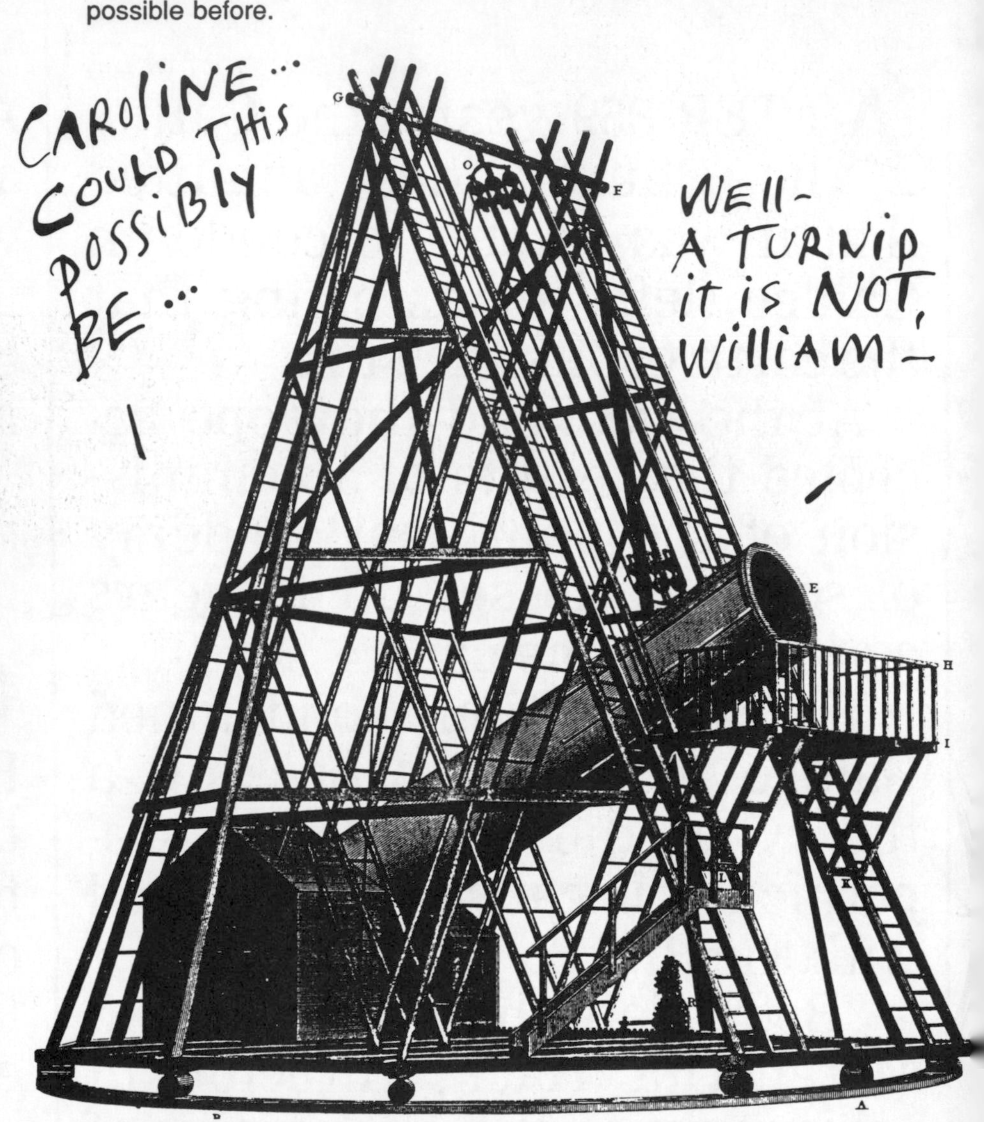


The planet had been recorded a score of times before by various astronomers, but none had noticed its distinctive appearance.

I called the planet *Georgium Sidus* after George III, King of England. The King made me his private astronomer, at a salary of £200 a year, and later added another £50 for my sister Caroline, who helped me with my work.

NEVER MIND THE HALF DOZEN COMETS I'VE DISCOVERED.... SILLY ME!

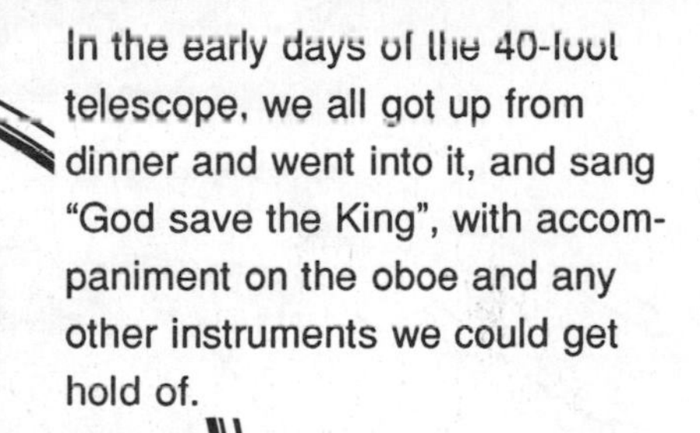

In the early days of the 40-foot telescope, we all got up from dinner and went into it, and sang "God save the King", with accompaniment on the oboe and any other instruments we could get hold of.

Just the same, everybody calls the planet Uranus, and not Georgium Sidus.

The development of astronomy continued in the nineteenth century, but it was not clearly understood by astronomers until early in the twentieth century that

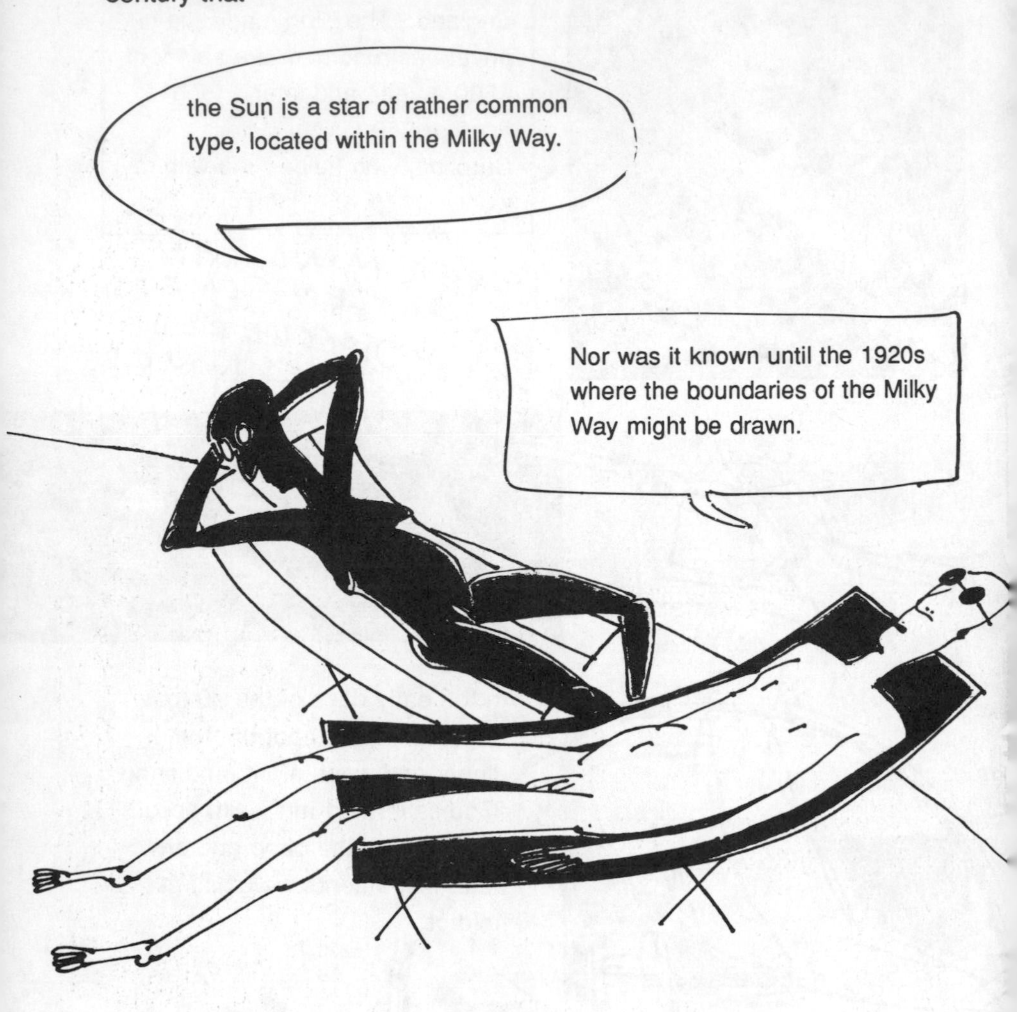


The spiral nebulae had not been definitely located beyond the confines of the Milky Way, nor had individual stars been distinguished in them.

Without reason to prefer one galaxy above another, one could no longer appeal to astronomy to prescribe a centre for the Universe.

# The Milky Way

The Milky Way, the Galaxy in which the Sun is a star, is visible to the naked eye as a long bright cloud in the sky. Galileo discovered in 1610 that it is made up of a great number of stars. There are several times $10^{11}$ stars in the Galaxy, which is shaped rather like a fried egg sitting in a cloud of smoke. Its mass is about $10^{12}$ times the mass of the Sun.

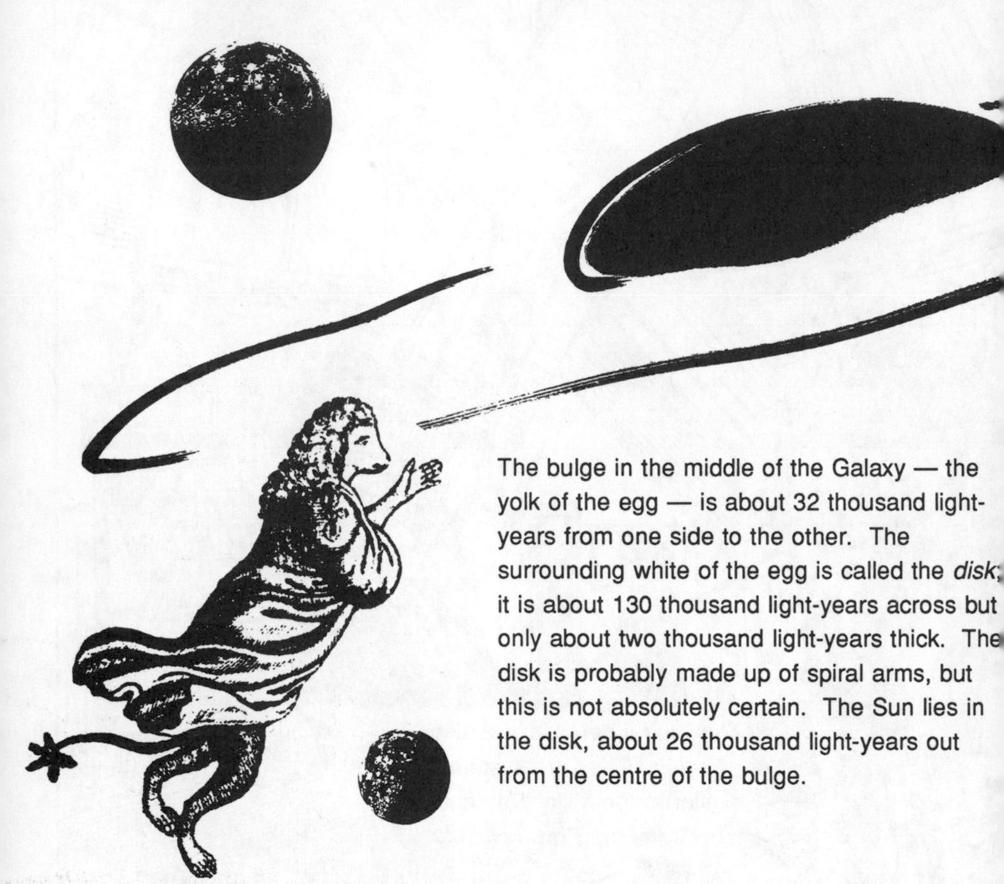

The bulge in the middle of the Galaxy — the yolk of the egg — is about 32 thousand light-years from one side to the other. The surrounding white of the egg is called the *disk*; it is about 130 thousand light-years across but only about two thousand light-years thick. The disk is probably made up of spiral arms, but this is not absolutely certain. The Sun lies in the disk, about 26 thousand light-years out from the centre of the bulge.

The cloud-of-smoke shape is called the *halo*. Not much of it can be seen but it may be five or six hundred thousand light-years from one side to the other. There are stars in the halo, but it is not known what the rest of it is made of.

The stars in the Galaxy appear to be rotating around its centre; it takes the Sun about 250 million years to go once around.

What's between the stars?

Gas and dust. Mostly hydrogen. In this part of the Galaxy between 1 and $10^8$ hydrogen atoms per litre.

How many atoms in a litre of water?

About $10^{26}$.

There may be from about a million to about $10^{13}$ stars in a single galaxy.

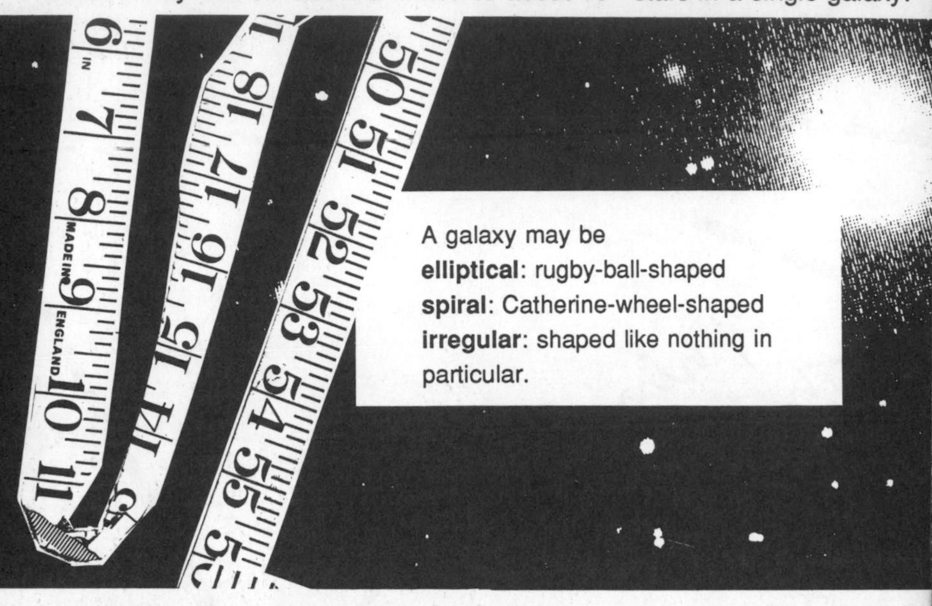

A galaxy may be
**elliptical**: rugby-ball-shaped
**spiral**: Catherine-wheel-shaped
**irregular**: shaped like nothing in particular.

Most galaxies are found not by themselves but in groups or *clusters* of tens, hundreds or thousands. The Milky Way belongs to a group called the Local Group, with at least 25 members.

The Local Group is about 3 million light years across.

The nearest big cluster, called the Virgo cluster (because it is seen in the same direction in the sky as the star constellation Virgo) is probably between 50 and 70 million light-years away. Distances to remote objects are very hard to determine with accuracy.

The Virgo cluster is thought to be about 7 million light-years in diameter, and has over 1200 known member galaxies.

The galaxies in each cluster appear to move around the cluster. Most clusters are grouped together. Both the Local Group and the Virgo cluster belong to the Local Supercluster, which is pancake-shaped, about 100 million light-years in diameter and 10 million light-years thick. It has about 100 member clusters.

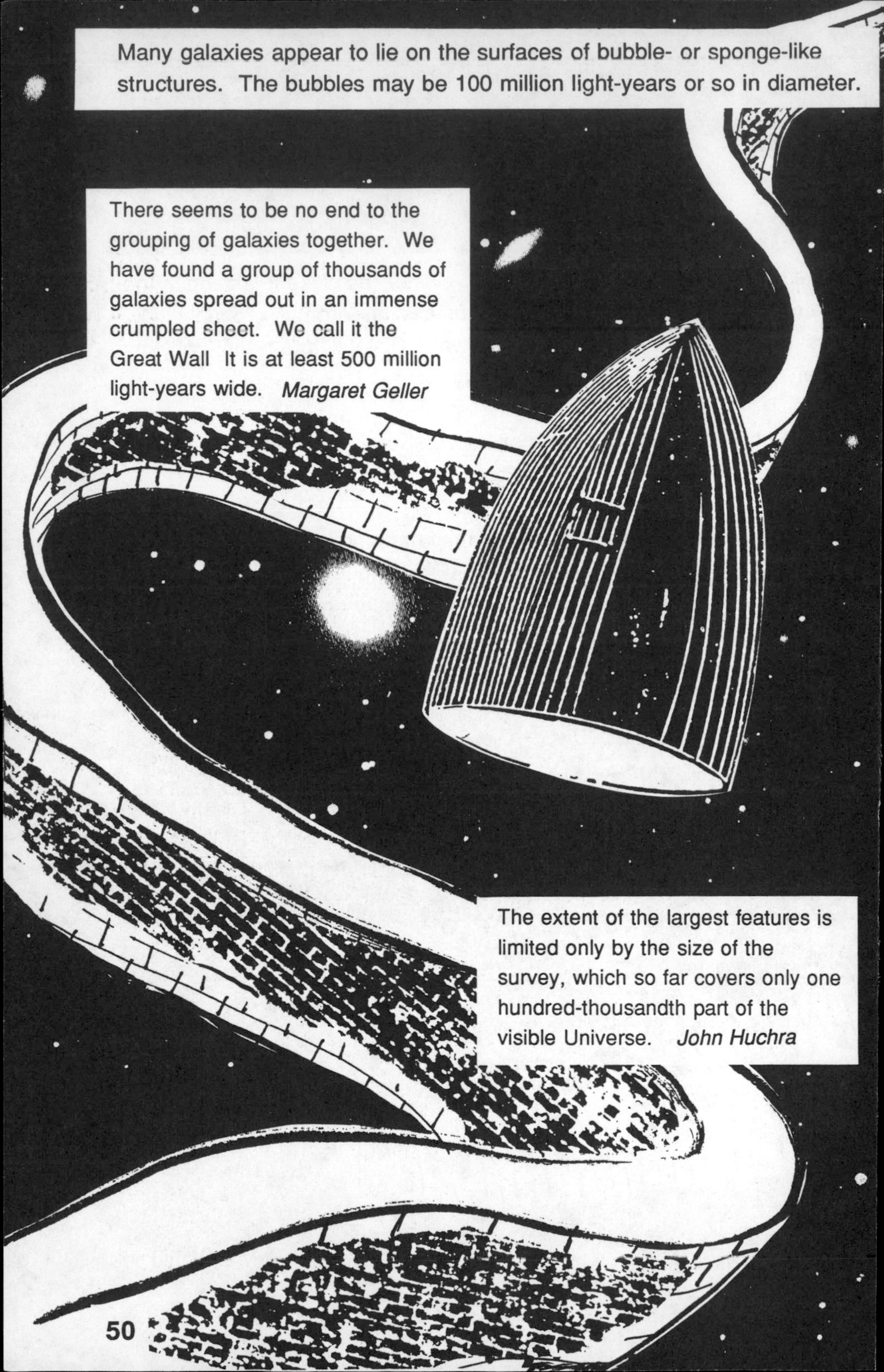
Many galaxies appear to lie on the surfaces of bubble- or sponge-like structures. The bubbles may be 100 million light-years or so in diameter.
There seems to be no end to the grouping of galaxies together. We have found a group of thousands of galaxies spread out in an immense crumpled sheet. We call it the Great Wall It is at least 500 million light-years wide. *Margaret Geller*
The extent of the largest features is limited only by the size of the survey, which so far covers only one hundred-thousandth part of the visible Universe. *John Huchra*

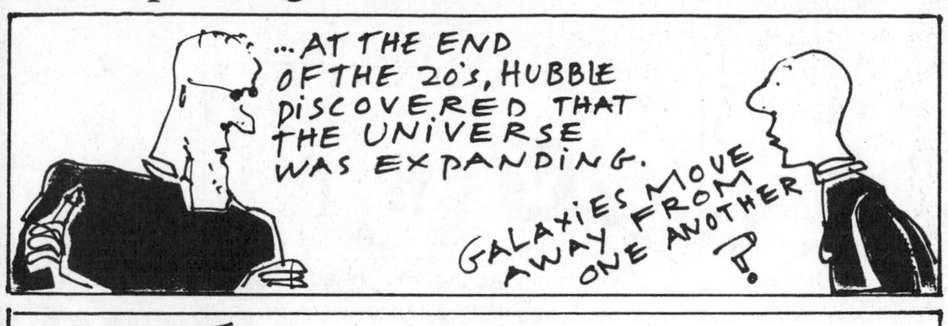
...AT THE END OF THE 20's, HUBBLE DISCOVERED THAT THE UNIVERSE WAS EXPANDING.
GALAXIES MOVE AWAY FROM ONE ANOTHER ?

THE MORE REMOTE THEY ARE FROM EACH OTHER...
THE FASTER THEY MOVE AWAY!

SO - THE MUTUAL RECESSION OF GALAXIES IMPLIES THAT THE MILKY WAY IS IN THE CENTRE ?!
THERE IS NO CENTER. THE APPEARANCE OF MUTUAL RECESSION WOULD BE...
...ENTIRELY SIMILAR FROM THE POINT OF VIEW OF AN OBSERVER FROM ANOTHER GALAXY.

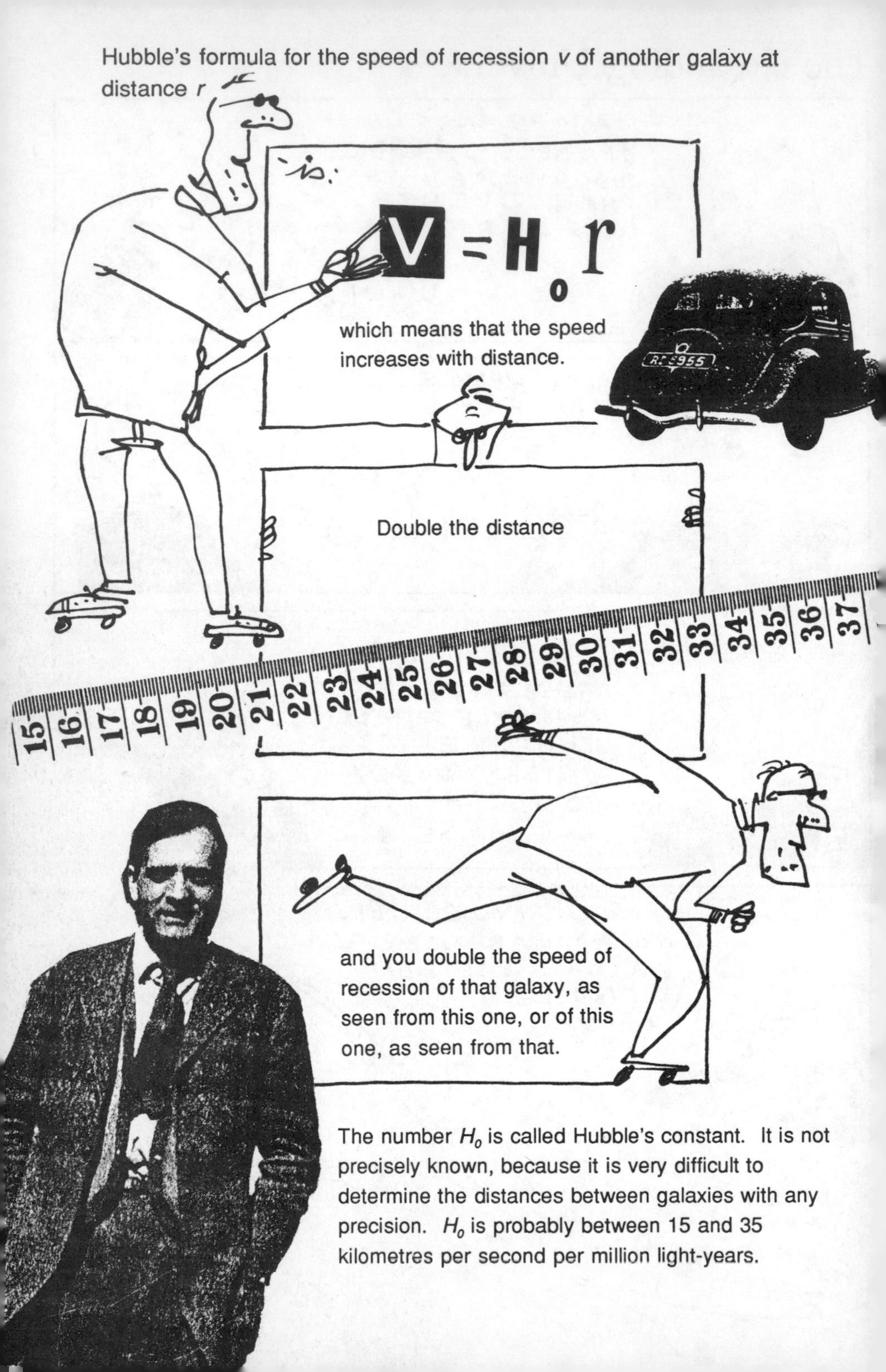

Hubble's formula for the speed of recession *v* of another galaxy at distance *r*
is:
$$V = H_0 r$$
which means that the speed increases with distance.
Double the distance
15 16 17 18 19 20 21 22 23 24 25 26 27 28 29 30 31 32 33 34 35 36 37
and you double the speed of recession of that galaxy, as seen from this one, or of this one, as seen from that.
The number $H_0$ is called Hubble's constant. It is not precisely known, because it is very difficult to determine the distances between galaxies with any precision. $H_0$ is probably between 15 and 35 kilometres per second per million light-years.

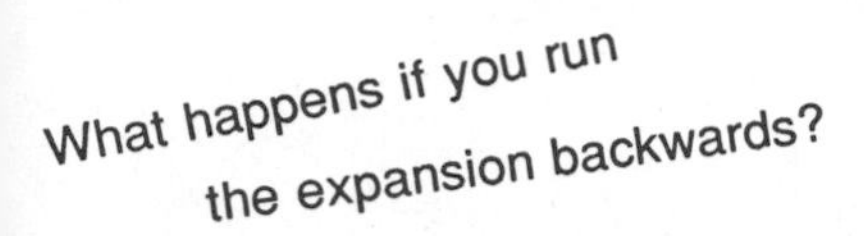


The galaxies must have been closer to one another in the past than they are now. At one time, their contents must have been squashed close together and the appearance of the Universe must have been very different.

If the laws of nature which are believed to apply now have always applied to the Universe, then there were initially no stars or galaxies but only a rapidly expanding hot gas of elementary particles — hot because it was compressed — out of which stars, galaxies and other structures subsequently developed. The further back you go, the hotter and denser do the contents of the Universe become. At least that is the majority view.

One idea was that the Universe started its life a finite time ago in a single huge explosion and that the present expansion is a relic of the violence of this explosion.

But the name *BIG BANG* caught on.

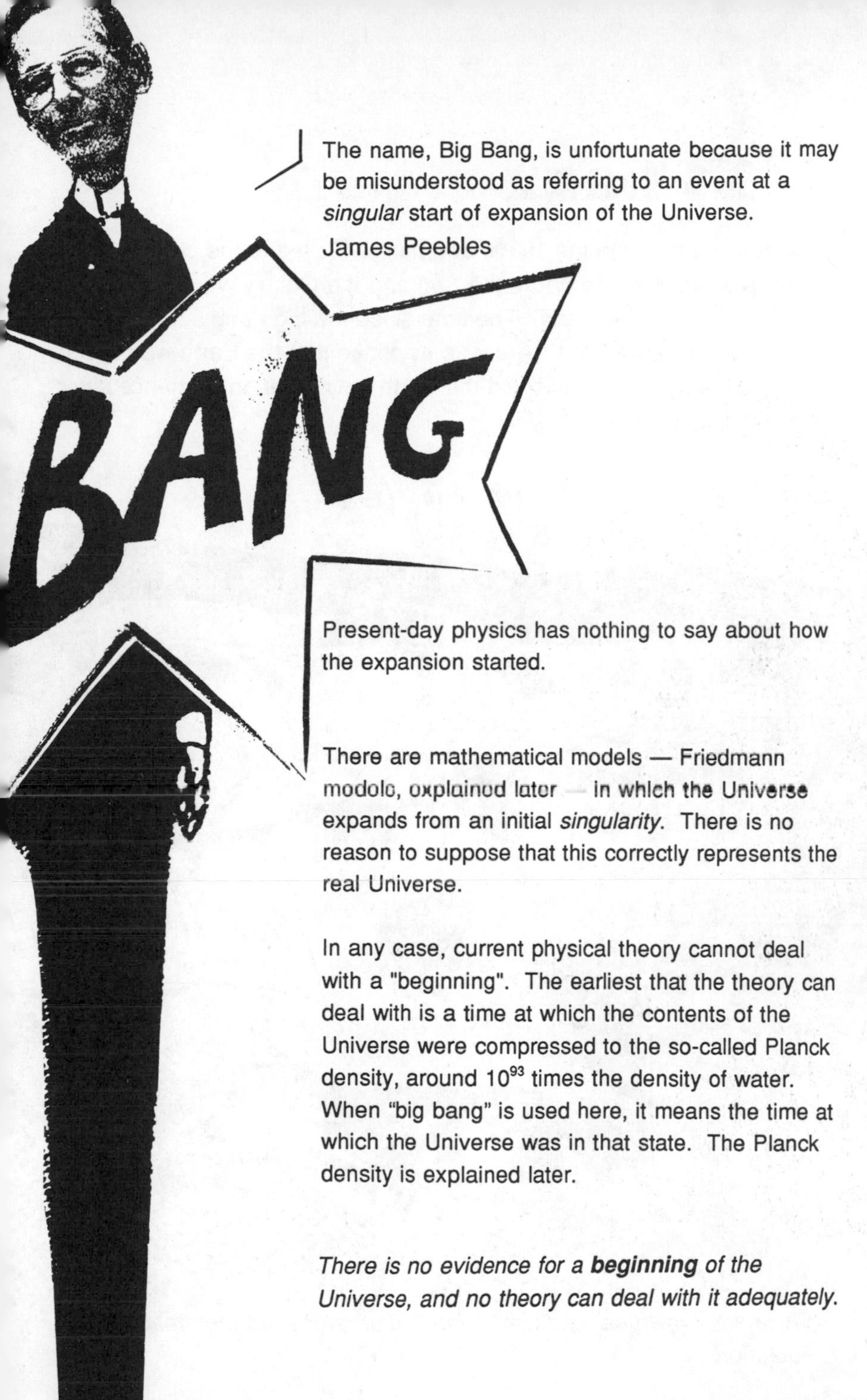


The name, Big Bang, is unfortunate because it may be misunderstood as referring to an event at a *singular* start of expansion of the Universe.
James Peebles

Present-day physics has nothing to say about how the expansion started.

There are mathematical models — Friedmann models, explained later — in which the Universe expands from an initial *singularity*. There is no reason to suppose that this correctly represents the real Universe.

In any case, current physical theory cannot deal with a "beginning". The earliest that the theory can deal with is a time at which the contents of the Universe were compressed to the so-called Planck density, around $10^{93}$ times the density of water. When "big bang" is used here, it means the time at which the Universe was in that state. The Planck density is explained later.

*There is no evidence for a* ***beginning*** *of the Universe, and no theory can deal with it adequately.*

The big bang idea ran into a difficulty.

Observations made in the 1930s and the 1940s led to the conclusion that the present ages of the Earth, the Sun and the Galaxy were greater than the time since the big bang. The time since the big bang seemed to be about 2 billion years. But there was evidence that the Earth was about 4½ billion years old. How could the Earth have been in existence since before the big bang?

BUT ONE DAY AT THE MOVIES –

A group known as the Cambridge Circus proposed the steady state solution.

In the steady state model of the Universe, the Universe expands, and new matter is created in the gaps left by the expansion at just such a rate as to maintain a roughly uniform distribution of matter, "steady" — of unchanging average density — forever.

The details change continually, but the overall appearance of the Universe is always the same.

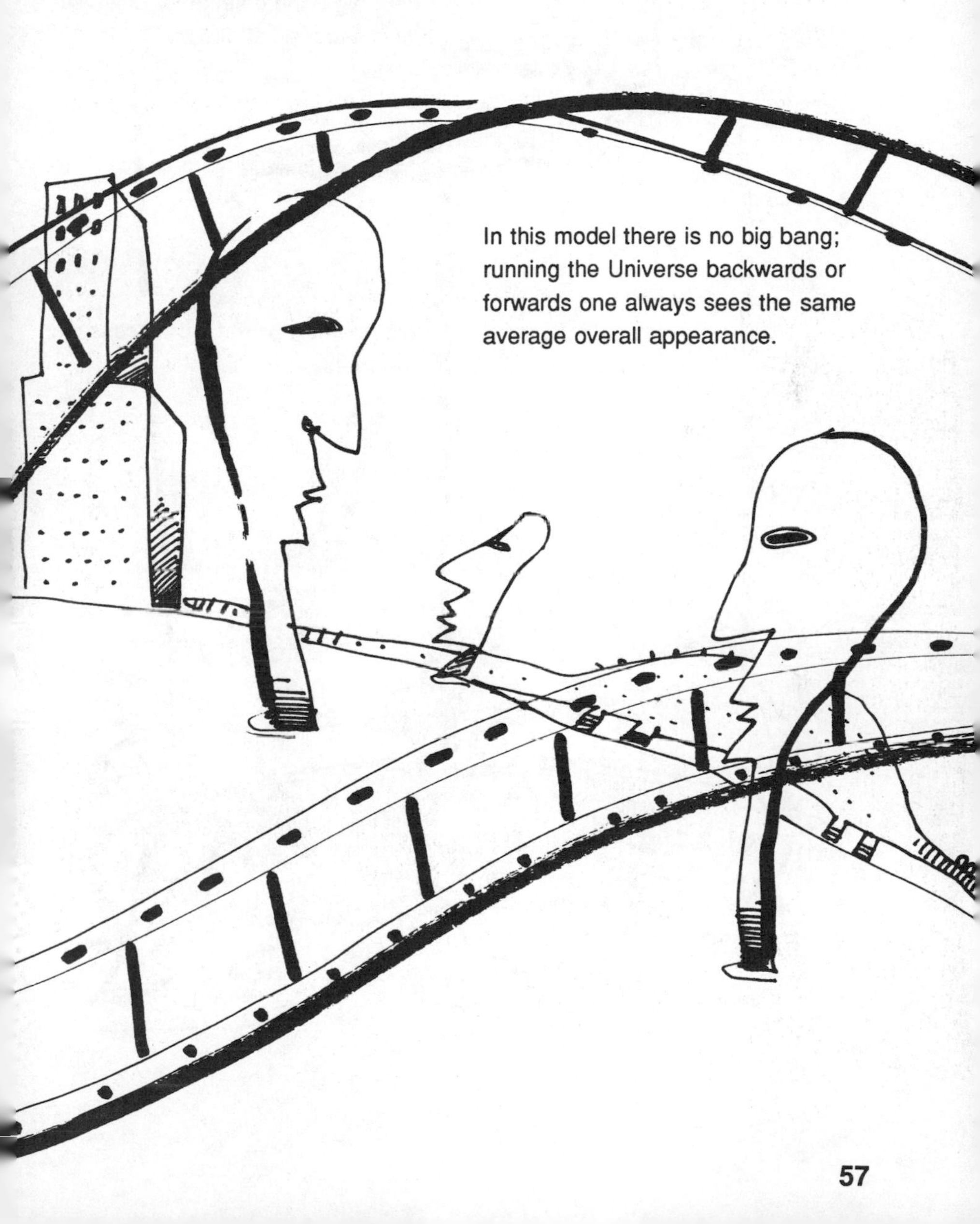

# The Cambridge Circus

We argued that continual creation of matter in small quantities in an always existing Universe...

The difficulty about the time scale persisted until 1952, when Walter Baade, working at Mount Palomar Observatory, revised the scale of distances to other galaxies.

Baade: the time since the big bang is likely to have been greater than five billion years — long enough for the evolution of Galaxy, Sun and Earth to have taken place.

Interest in the steady state model declined, and almost vanished after the discovery in 1965 of the cosmic background radiation …

…EXPLAINED LATER

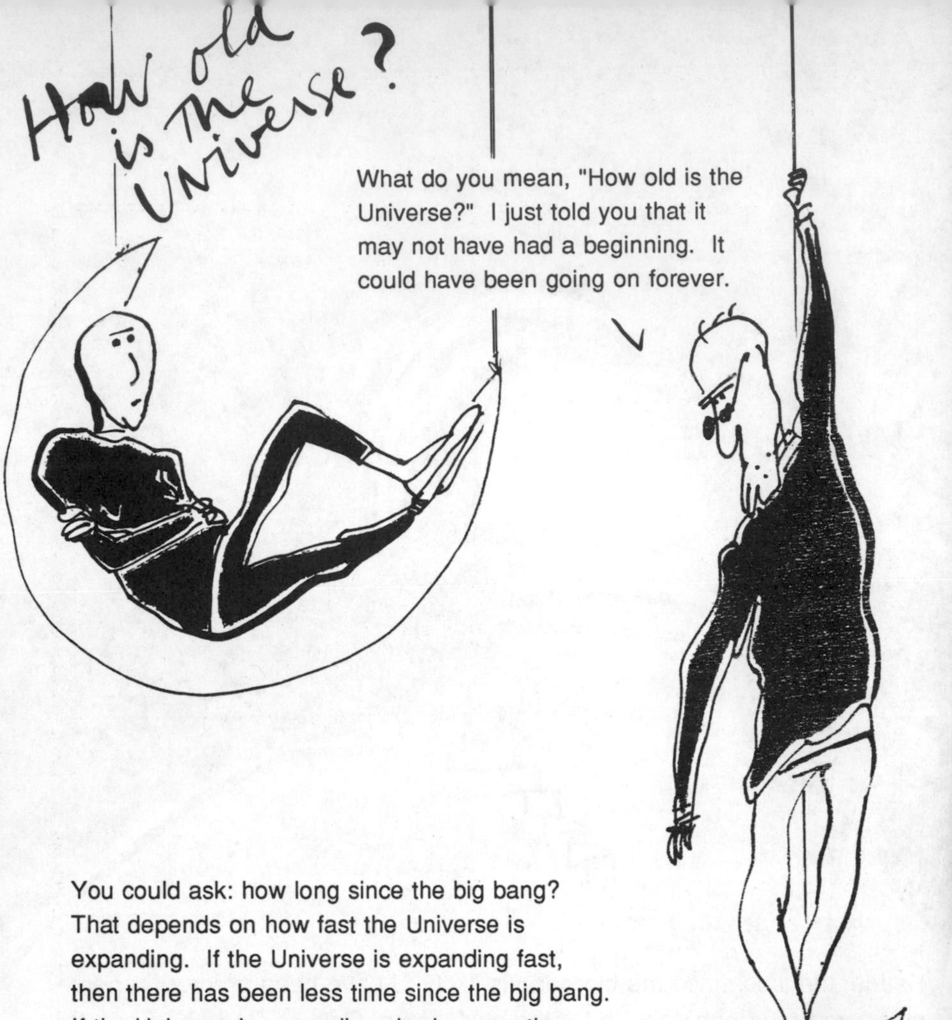


You could ask: how long since the big bang? That depends on how fast the Universe is expanding. If the Universe is expanding fast, then there has been less time since the big bang. If the Universe is expanding slowly, more time. Astronomers agree on how fast the other galaxies move away, but they can't agree about how far away the other galaxies are...

So they can't tell exactly how long it is since the big bang. Probably between 10 and 20 billion years. Which is about the same as estimates of the age of the Milky Way. The time-scale difficulty might come back to life one of these days.

In 1965, Arno Penzias and Robert Wilson, two radio astronomers working at Bell Laboratories in New Jersey, were tracing sources of radio noise.

They had to clean the pigeon droppings out of the antenna first.

Since the work of Penzias and Wilson, almost all cosmologists now accept the idea of the big bang, meaning an early, highly condensed state from which the expansion of the Universe may be supposed to have proceeded. Remember that "big bang" does not mean the beginning of the Universe. Physics has nothing to say about a "beginning".

# Laws of nature

Modern cosmology begins with some basic assumptions.

*The rest of the Universe is made of atoms of the same kind as those on the Earth.*

*The same laws of nature apply in the rest of the Universe as on the Earth.*

*The same laws apply to the past as to the present.*

The idea of a 'law of nature' goes back at least to the seventeenth century French philosopher René Descartes.

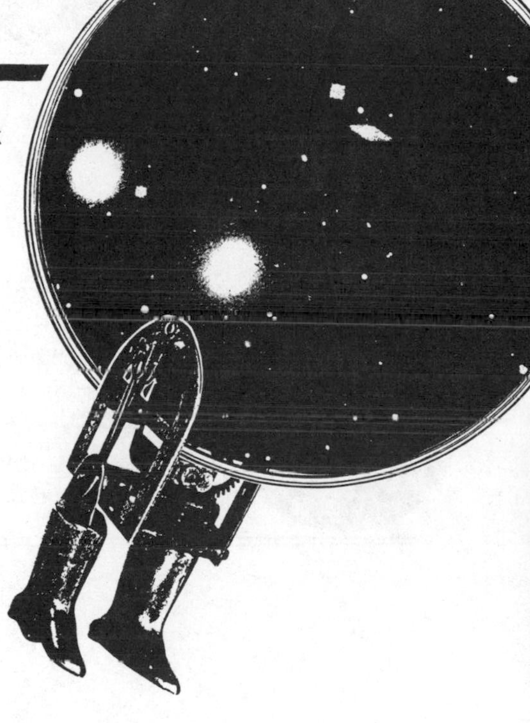

I suppose that God rules the Universe entirely by 'laws of nature' which were decided upon at the beginning.

In contrast, during the middle ages God participated in the day-to-day running of the Universe.

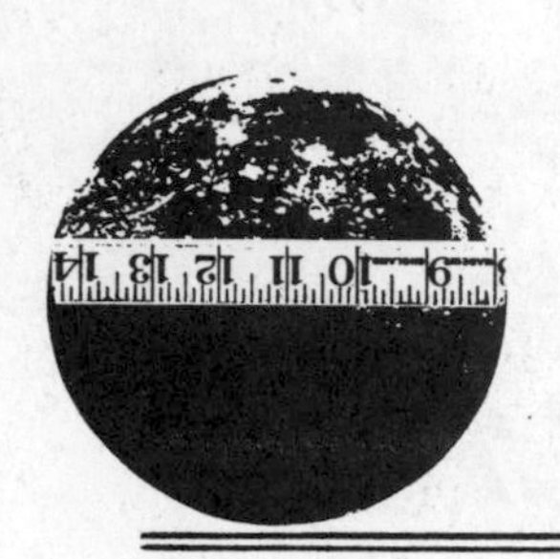

Perhaps the idea of specifically quantitative 'laws of nature' came from the international associations of merchants, like the old Hansa League, who had laws of their own, which dealt with numbers, weights and measures.

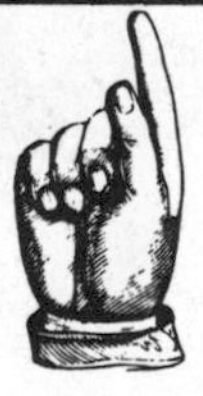

***Eugene Wigner*: It is not at all natural that "laws of nature" exist, much less that man is able to discover them.**

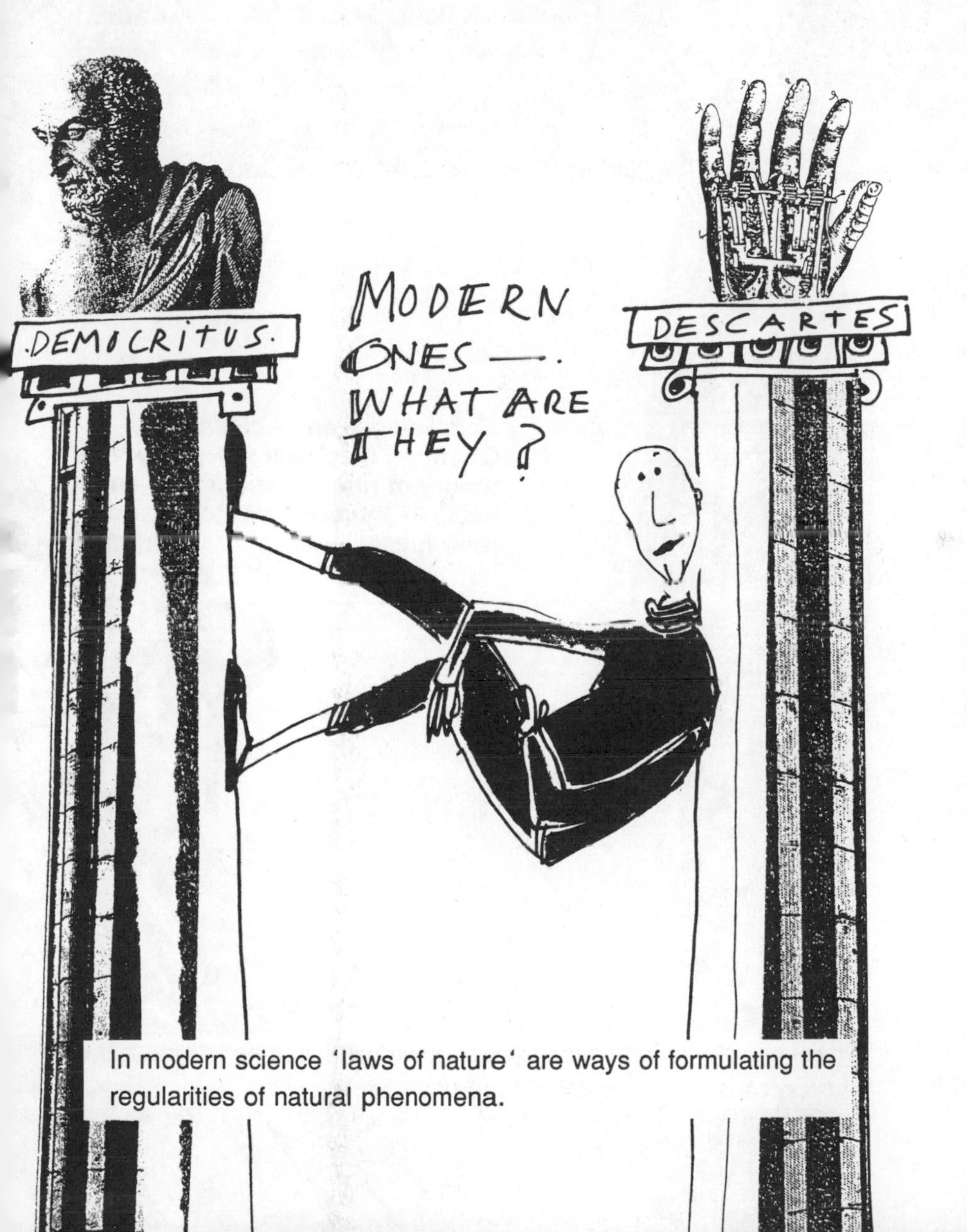


In modern science 'laws of nature' are ways of formulating the regularities of natural phenomena.

# The Laws of Modern Physics

The relevant laws are laws of physics, embodied in two fundamental theories which describe the behaviour of matter and radiation: general relativity and quantum mechanics.

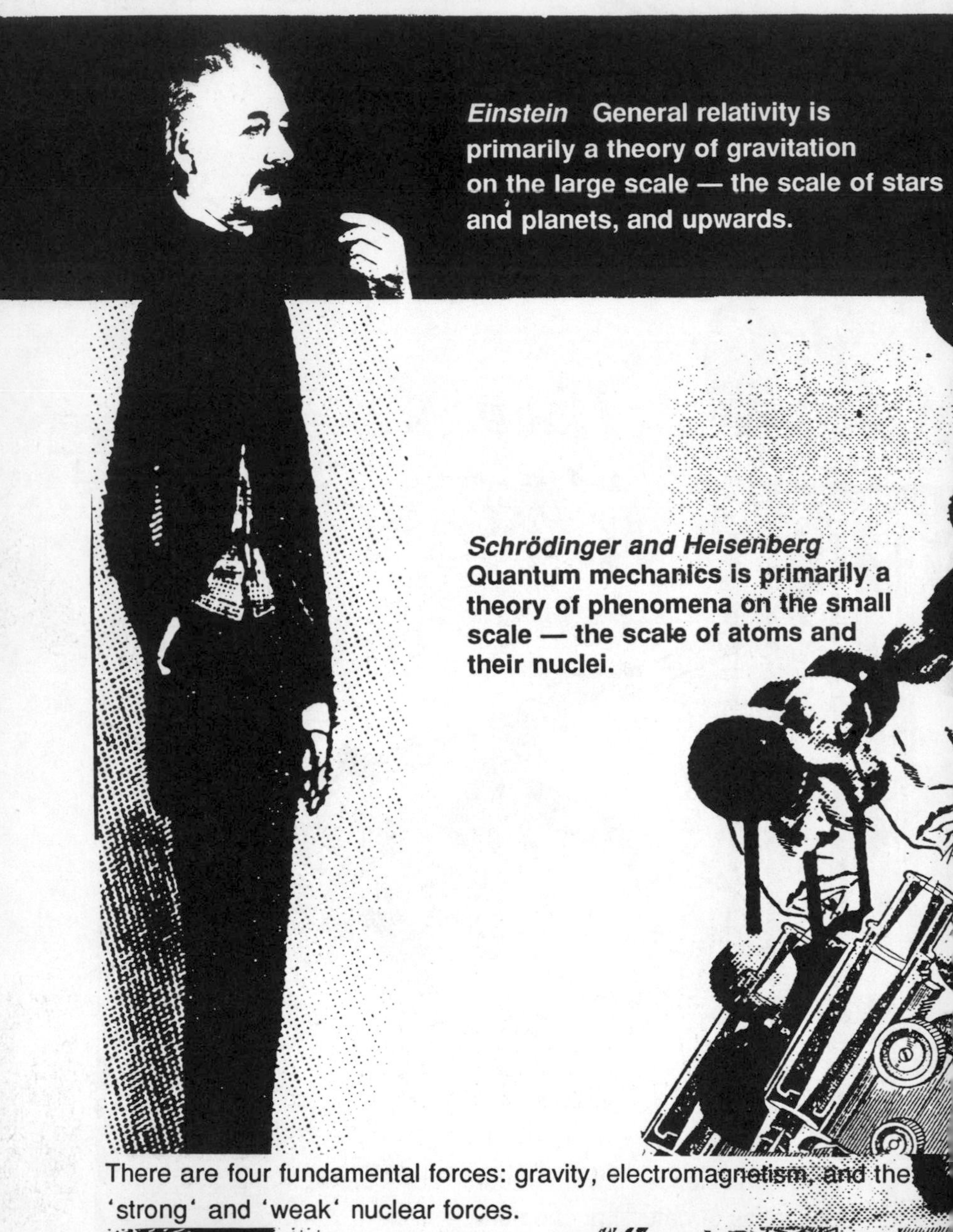

*Einstein* General relativity is primarily a theory of gravitation on the large scale — the scale of stars and planets, and upwards.

***Schrödinger and Heisenberg***
**Quantum mechanics is primarily a theory of phenomena on the small scale — the scale of atoms and their nuclei.**

There are four fundamental forces: gravity, electromagnetism, and the 'strong' and 'weak' nuclear forces.

Energy is conserved: it may be converted from one form to another, but there is no process known in which it is created or destroyed. However energy can manifest itself as mass, with the result that sometimes conversion of energy from one form to another may appear as the conversion of mass into energy or vice versa, according to Einstein's formula $E = mc^2$. This is the case when stars radiate energy, as you will see later.

Under the moderate conditions which prevail on the Earth, most matter exists in the form of atoms. Each atom has a central nucleus (the plural is nuclei), surrounded by a cloud of electrons. Nearly all the mass of the atom resides in the nucleus. Atoms combine to form molecules; in a pure chemical compound every molecule is made up of the same combination of atoms.

For example every molecule of water consists of two atoms of hydrogen and one of oxygen.

"Hydrogen" and "oxygen" are the names of chemical elements. To understand what a chemical element is, you need to know something about the structure of atoms.

The forces which hold the electrons in an atom are electromagnetic forces. Responsiveness to electromagnetic forces is measured by electric charge, which can be positive or negative. The nucleus of an atom is charged, and so are the surrounding electrons. By convention, the charge on the nucleus is called positive, while the charge on an electron is called negative.

Opposite charges attract, like charges repel —

To begin with, one may think of an atomic nucleus as made up of protons and neutrons (protons and neutrons are thought to be made up of still more basic particles called quarks). A conventional unit of electric charge is the charge on the proton: a proton has charge +1. An electron has charge -1, and a neutron has no charge. An atom by itself — a neutral atom — has zero total charge. The number of electrons in the cloud around the nucleus is the same as the number of protons in the nucleus itself, which is called the atomic number.

# ELEMENTS

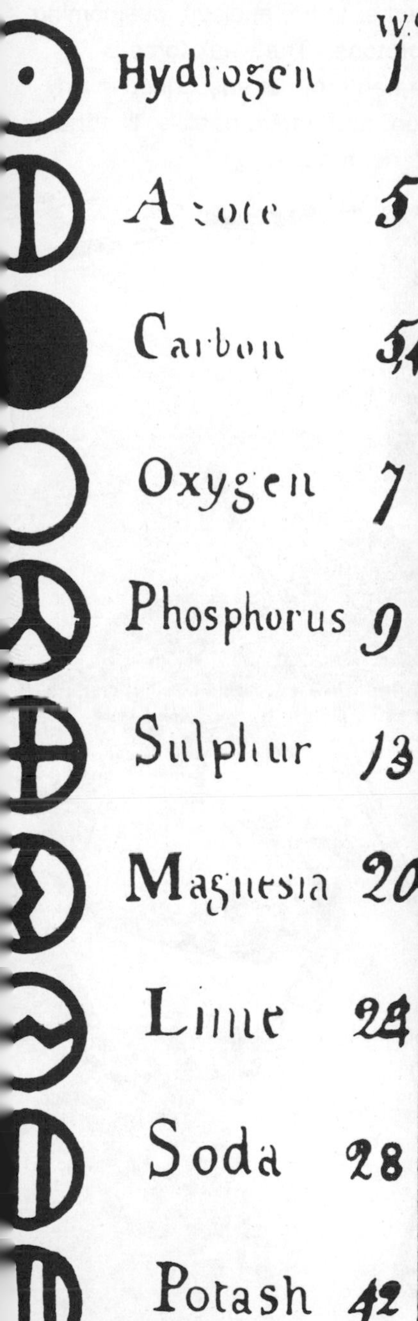


Atoms which have the same atomic number — the same number of protons in their nuclei — are atoms of the same element. The chemical behaviour of an atom is determined almost completely by the atomic number. Atoms with atomic number 1 are atoms of hydrogen, those with atomic number 2 are helium, ... those with atomic number 92 are uranium.

Nuclei of a particular element need not all have the same number of neutrons. For example helium nuclei may have one neutron or two, as well as the two protons which they all have. An isotope of an element is a particular kind of nucleus of that element with a particular number of neutrons.

Protons and neutrons are called nucleons. The number of nucleons in a nucleus of a particular isotope is called the nucleon number. Ordinary hydrogen has one proton and no neutrons; its nucleon number is 1. There is a hydrogen isotope called deuterium, which has one proton and one neutron, so that its nucleon number is 2. The nucleon number is often written as a superscript to the conventional abbreviation of the name of an element: hydrogen is H, ordinary hydrogen is $^{1}$H, and deuterium is $^{2}$H. Hydrogen also has an isotope $^{3}$H, called tritium. Helium is abbreviated He, and the common isotope of helium is $^{4}$He.

The forces which act between nucleons are called nuclear forces.

Two different nuclear forces are recognised, called "strong" and "weak".

The strong force holds the nucleons together in the nucleus, overcoming the electrostatic repulsion between the protons. The weak force is involved in the decay of neutrons: a free neutron, not ensconced in any nucleus, decays into a proton, an electron, and a third particle, like the neutron not electrically charged, called a neutrino.

# Energy for stars and H-bombs

At sufficiently high temperatures, which occur naturally in the interiors of stars and are produced artificially in hydrogen bombs, light nuclei fuse together to form heavier ones. In a series of steps, four $^1H$ nuclei fuse to form a $^4He$ nucleus. The mass of a $^4He$ nucleus is slightly less than four times the mass of a $^1H$ nucleus.

**The "slightly less" is released as energy according to Einstein's formula $E = mc^2$.**

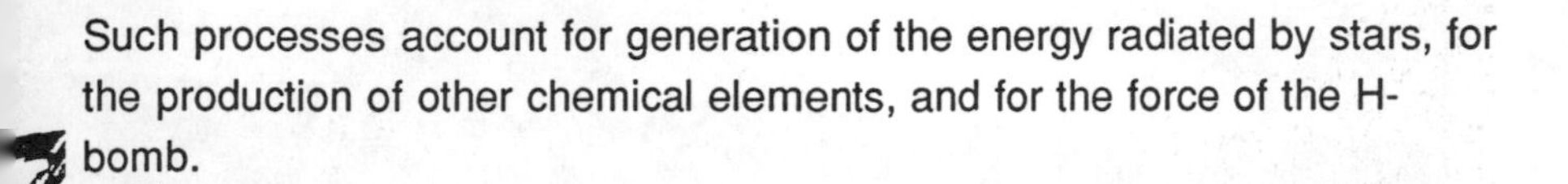

Such processes account for generation of the energy radiated by stars, for the production of other chemical elements, and for the force of the H-bomb.

Nearly all of what humans know of the rest of the Universe they know from reception of electromagnetic radiation.

*Visible light,*
*Radio waves,*
*Infra-red and ultraviolet radiation,*
*X-rays and gamma rays.*

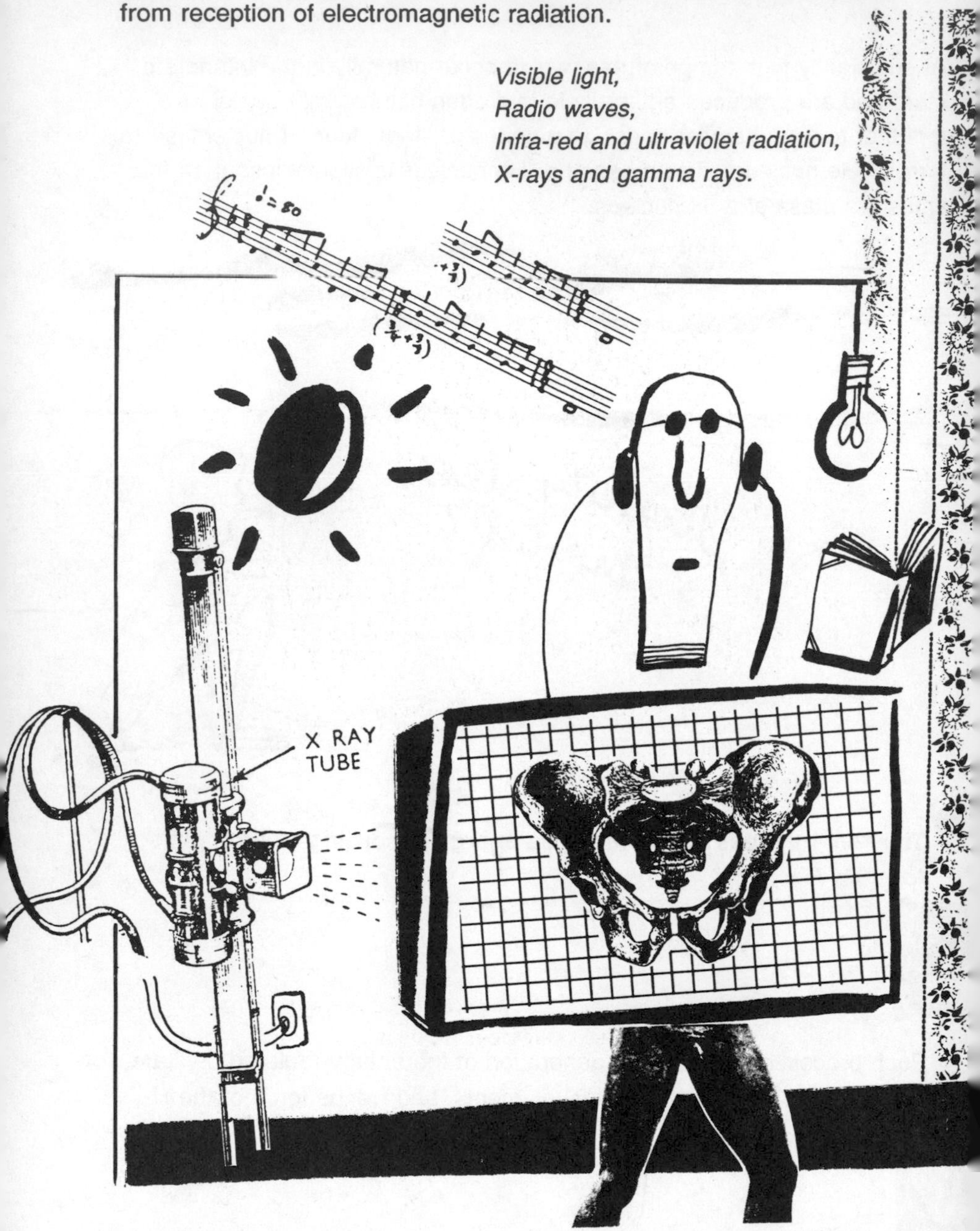

X RAY
TUBE

In the 1860s James Clerk Maxwell completed the unification of the theories of electricity and of magnetism, which had been in progress for about half a century.

The theory of electromagnetism describes the relations between electric and magnetic fields, and their production by electric charges and currents. The existence and properties of electromagnetic radiation, which is produced by moving charges, may be deduced from these equations:

$$\nabla \times \mathbf{E} + \frac{\partial \mathbf{B}}{\partial t} = 0$$

$$\nabla \times \mathbf{H} - \frac{\partial \mathbf{D}}{\partial t} = \mathbf{J}$$

$$\nabla \cdot \mathbf{D} = \rho$$

$$\nabla \cdot \mathbf{B} = 0$$

The different kinds of electromagnetic radiation may be distinguished by frequency. The radiation consists of oscillating electric and magnetic fields; the frequency is the rate of oscillation. It is measured in cycles per second, or Hertz (Hz), named after the physicist Heinrich Hertz.

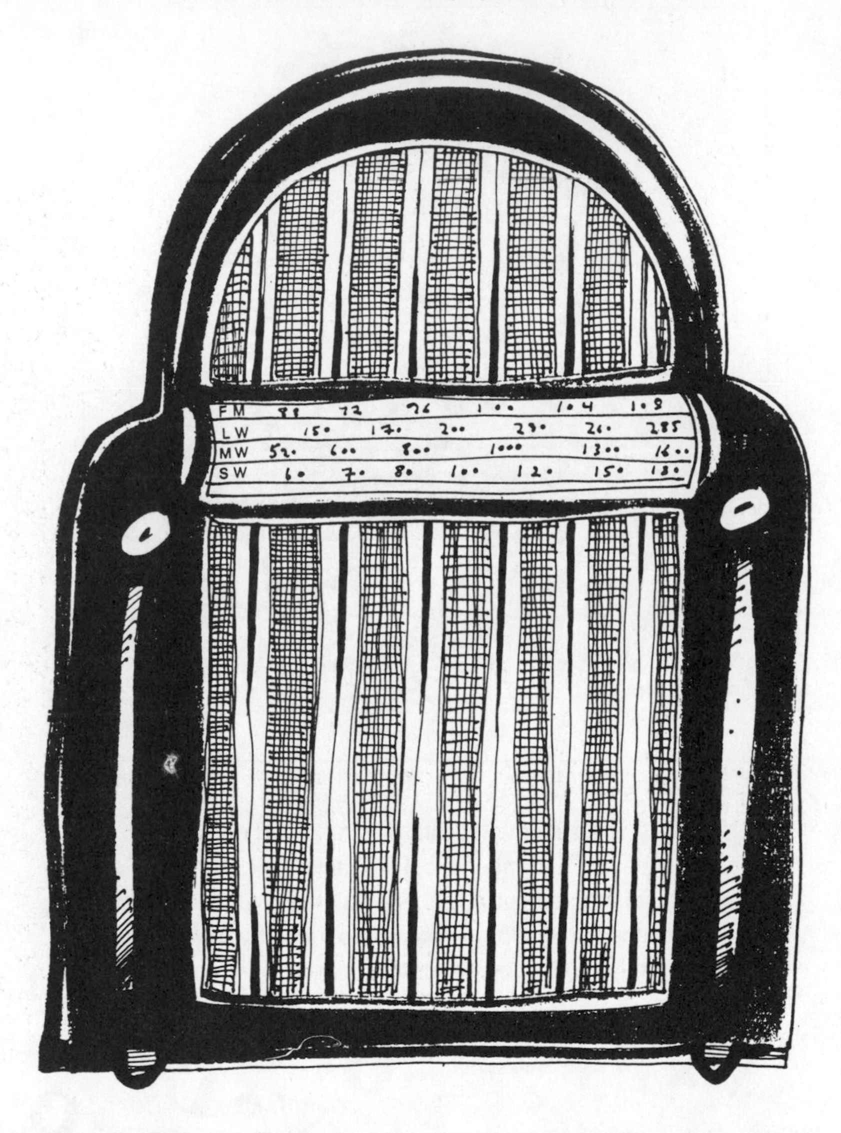


All electromagnetic radiation travels in empty space with the speed of light. In the Earth's atmosphere, the speed is slightly lower, in a denser medium, still lower.

A Megahertz (MHz) is a million Hertz. The radio waves from an FM station at 100 on the dial have a frequency of 100 MHz. To every frequency there corresponds a wavelength. 100 MHz radio waves have a wavelength of just under 3 metres.

# Waves or particles?

Depending on the circumstances, radiation may be perceived as waves, sharing some physical characteristics of sound waves and water waves...

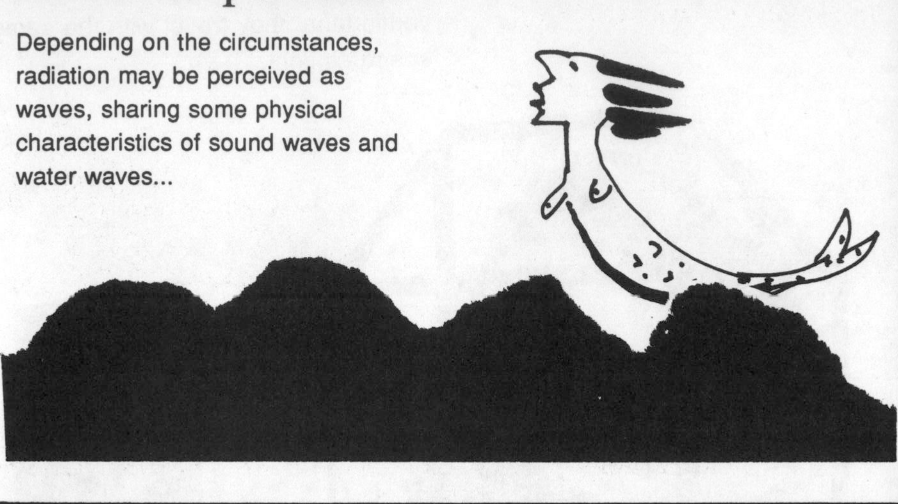

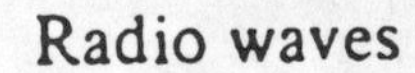


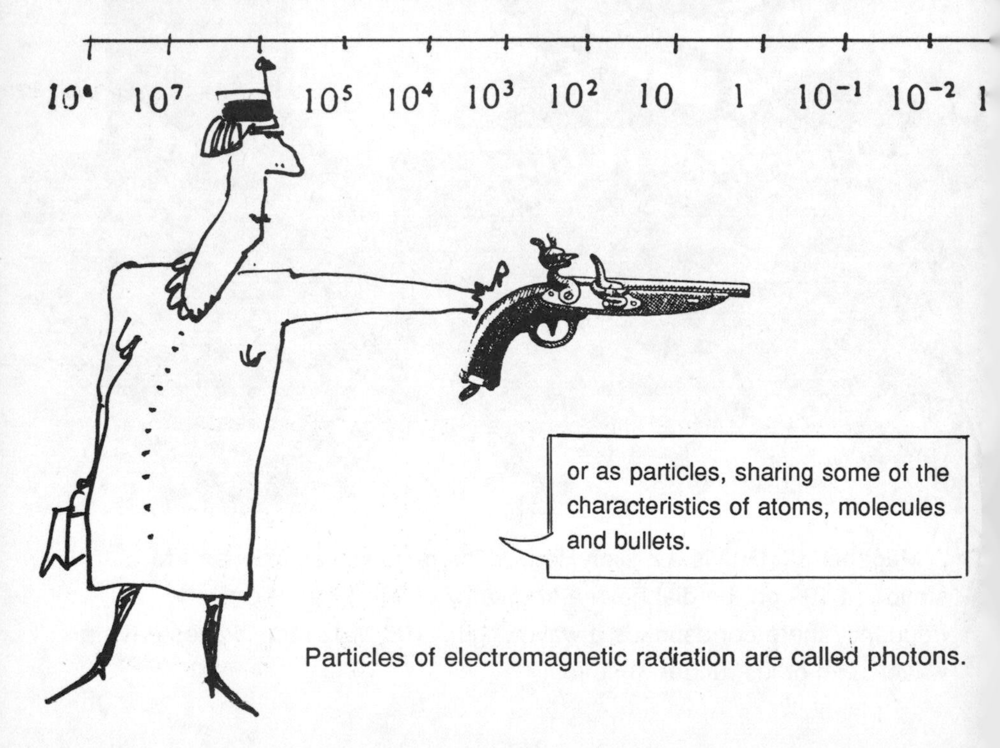


Particles of electromagnetic radiation are called photons.

Frequency and wavelength are related by the formula:
**frequency × wavelength = c.**

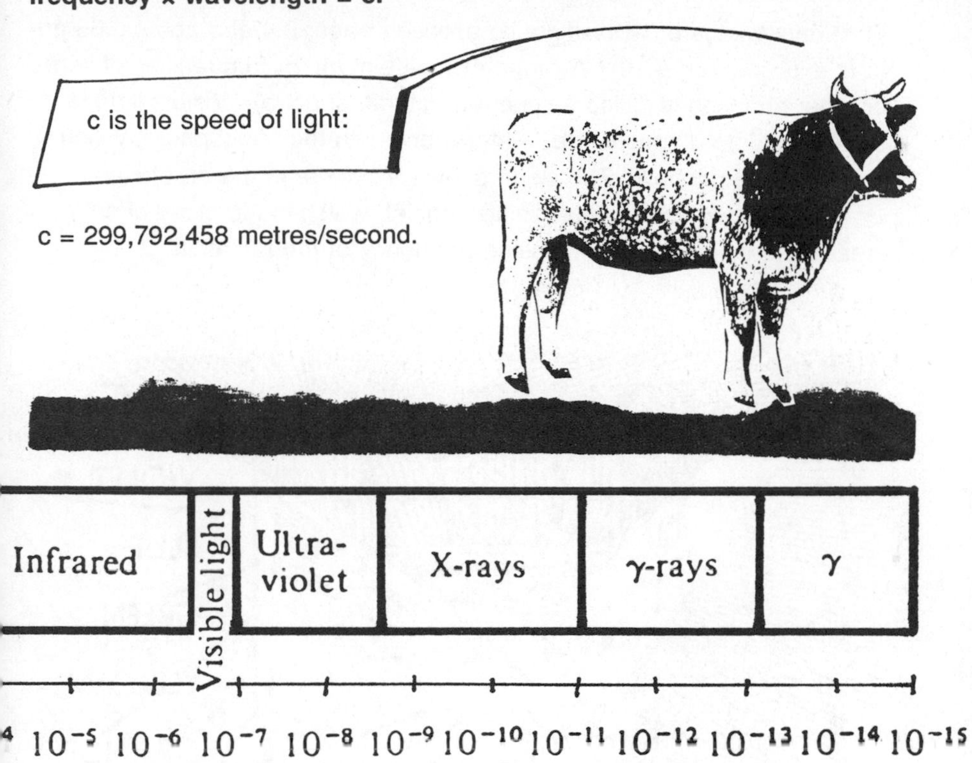


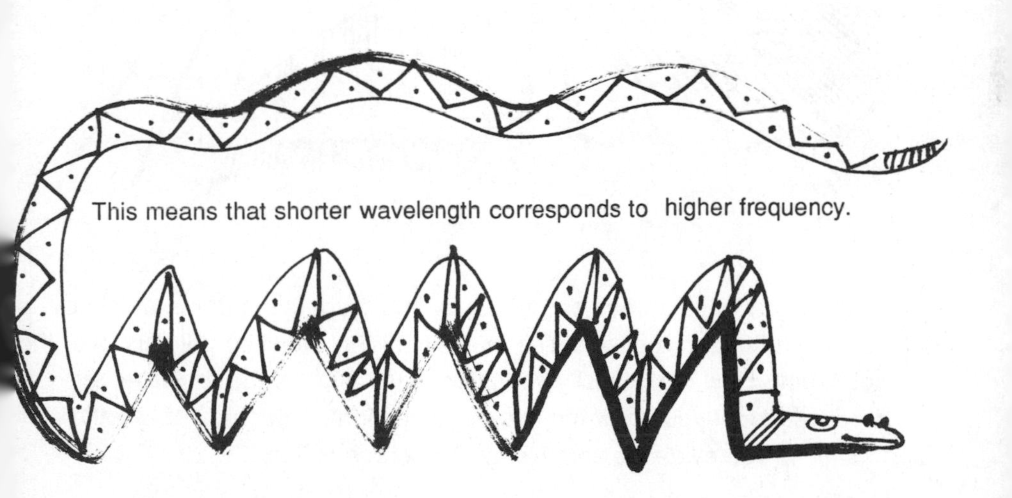

# Visible light

The wavelengths of visible light lie between about 0.4 and about 0.66 μm (μ is explained on p 10). A range of wavelengths or frequencies of light or other radiation is called a spectrum (plural: spectra). Visible light is spread out into a spectrum by water droplets in the atmosphere, which produces the rainbow. The colours always appear in the order Red, Orange, Yellow, Green, Blue, Indigo, Violet, which is the order of decreasing wavelength, or increasing frequency of the radiation.

Light may also be spread out artificially into a spectrum, for example by passing it through a prism. The intensity of the radiation will vary over the spectrum. How the intensity varies depends on how the radiation is produced, so if you know the intensity variation of a source of radiation, this gives you evidence about what the source is like.

# Spectra

The spectrum is often recorded on a photographic plate in the form of a long rectangular image crossed by light or dark lines corresponding to peaks or valleys of intensity at particular wavelengths.

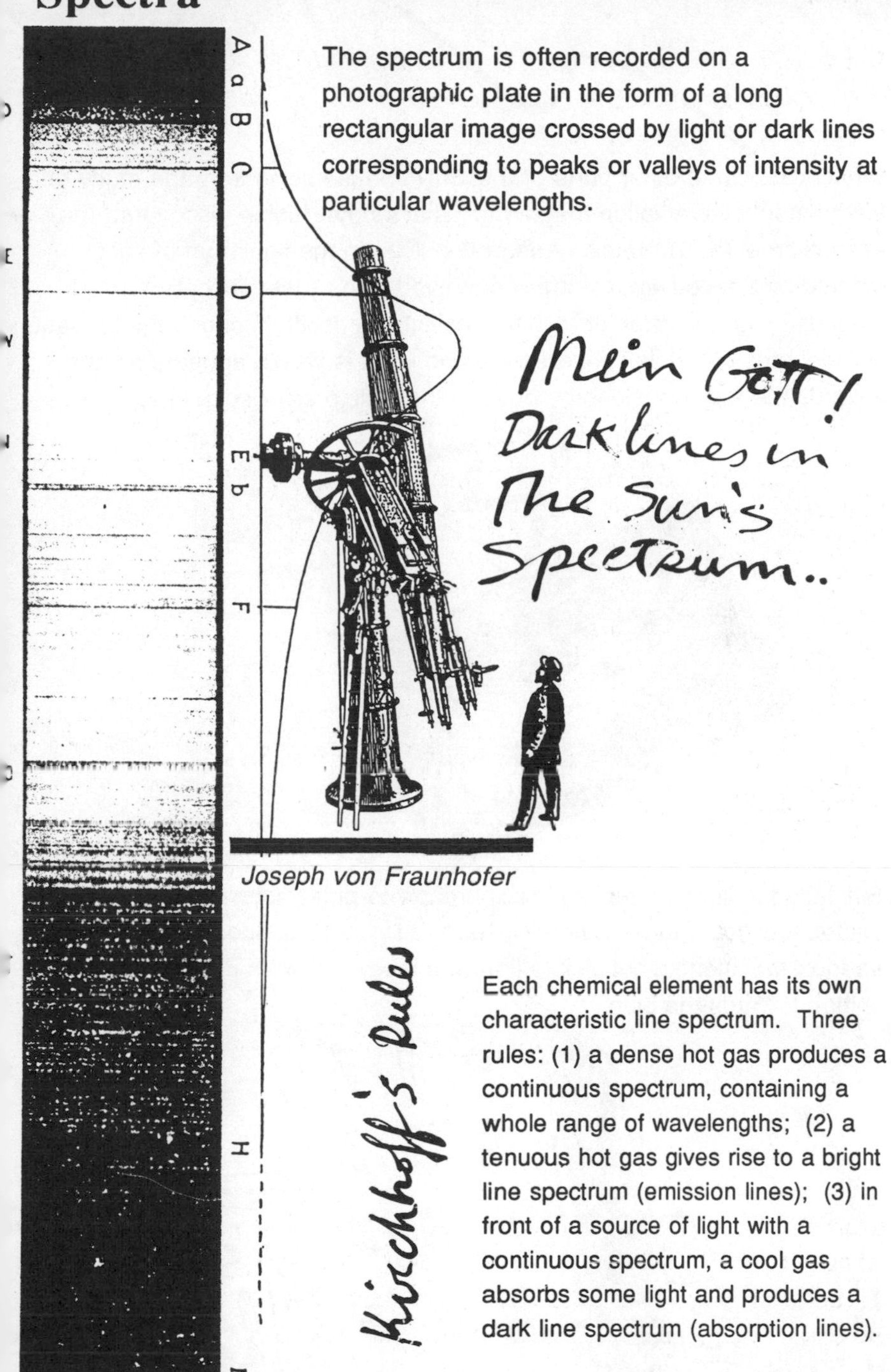


Joseph von Fraunhofer

Each chemical element has its own characteristic line spectrum. Three rules: (1) a dense hot gas produces a continuous spectrum, containing a whole range of wavelengths; (2) a tenuous hot gas gives rise to a bright line spectrum (emission lines); (3) in front of a source of light with a continuous spectrum, a cool gas absorbs some light and produces a dark line spectrum (absorption lines).

# Redshift — Blueshift

What is known of other parts of the Universe is known from the study of electromagnetic radiation they emit. That's how Hubble discovered the expansion of the Universe. Absorption lines in the spectrum of light emitted by a receding object are observed to be systematically shifted towards longer wavelengths; this is called a redshift, because the longest wavelengths of visible light are red, and what is visible appears shifted towards the red (wavelengths beyond the visible are also shifted).

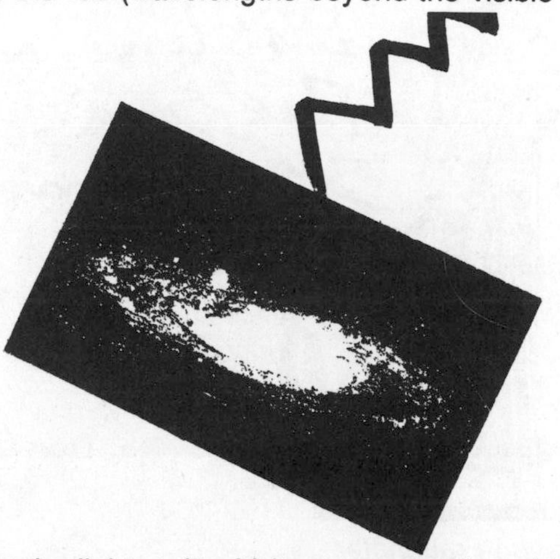

Similarly the light emitted by an approaching object is systematically shifted towards shorter wavelengths; this is called a blueshift, because the shortest wavelengths of visible light are blue, and what is visible appears shifted towards the blue.

In this respect light waves resemble sound waves: if the source of the waves is receding from you, the waves will seem to you to be longer than if the source is at rest. If the source of the waves is approaching you, the waves will seem to you to be shorter. In the case of sound waves it is this effect that produces the drop in pitch of an ambulance siren as the ambulance passes you.

# Special relativity

Towards the end of the nineteenth century it became apparent that Maxwell's electromagnetic theory could not be fitted into the same framework as Newton's mechanics.

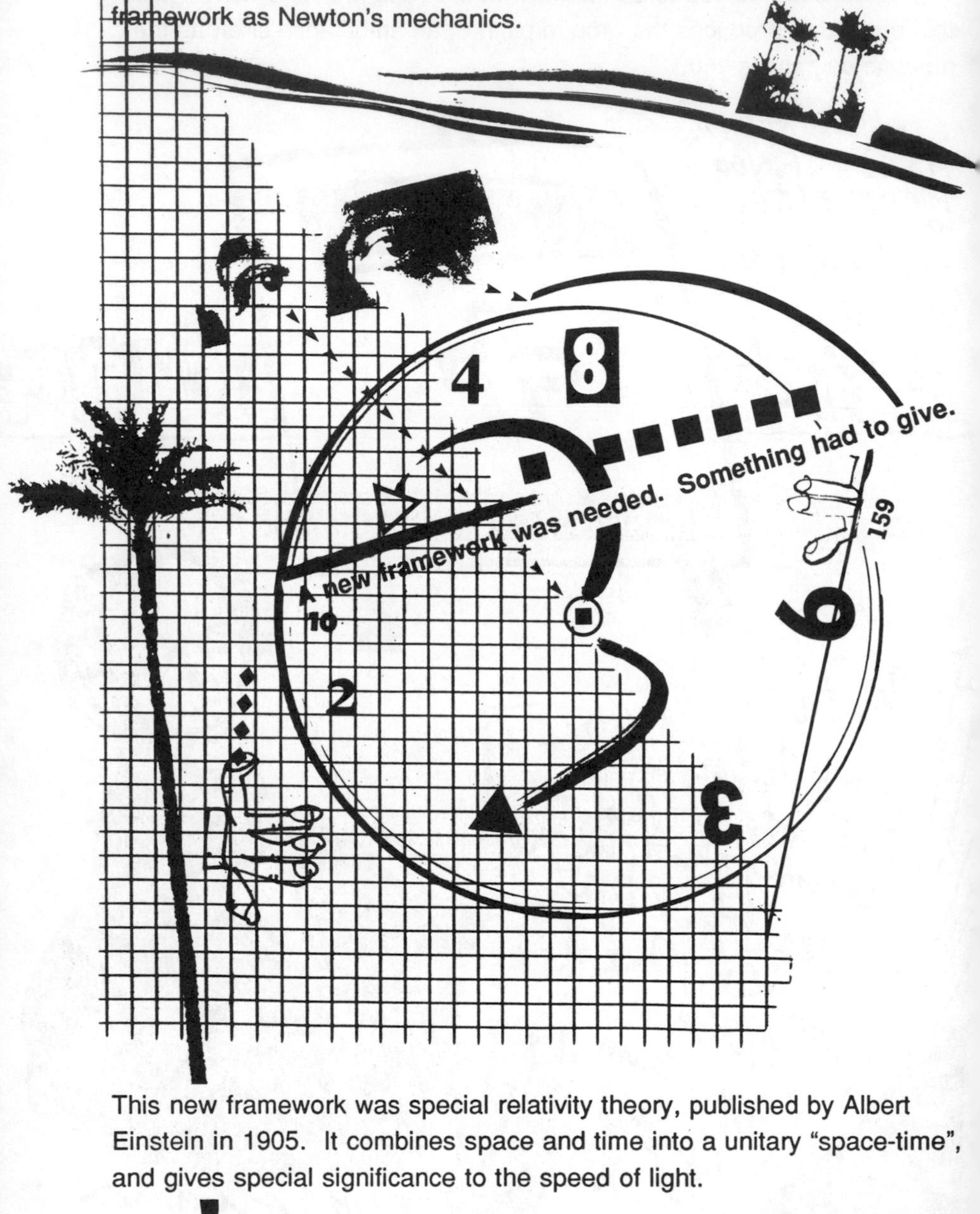

A new framework was needed. Something had to give.

This new framework was special relativity theory, published by Albert Einstein in 1905. It combines space and time into a unitary "space-time", and gives special significance to the speed of light.

# Space-time diagrams

1.

What are dimensions? Space has three dimensions: length, breadth and thickness.

At the Earth's surface North-South, East-West, and up-down.

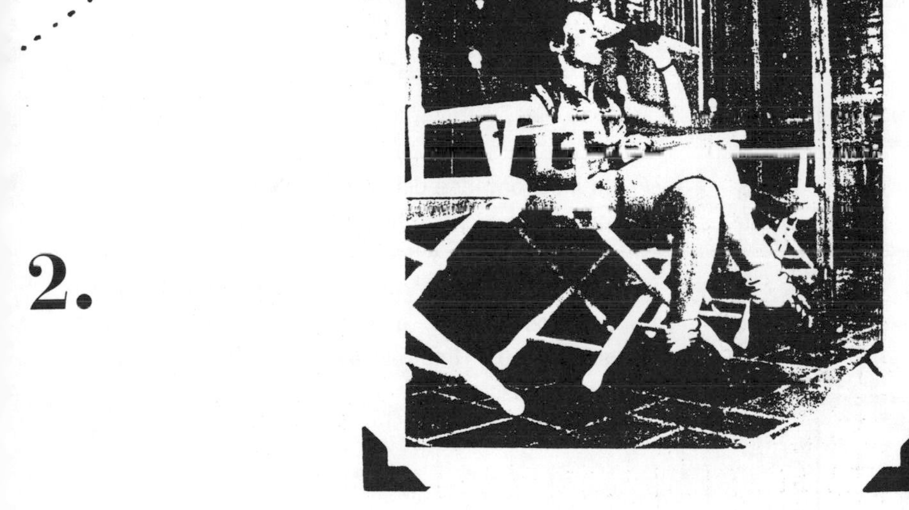

2.

A photograph is a two-dimensional representation of a three-dimensional scene. A map has two dimensions (the piece of paper on which the map is printed has three dimensions, but is very limited in one of them compared to the other two).

**3.**

Time also may be counted as a dimension. A space-time diagram shows positions in time as well as in space. In this diagram, there is one space dimension, which runs left and right. Here and in the diagrams which follow, time runs upward.

**4.**

In most of these diagrams, only one space dimension will be shown. Each person, or particle, or planet has a world-line in space-time. In the diagram, your world-line is just a graph showing your movements. If you stay in one place, it goes straight up. Time passes, but there is no change of space position.

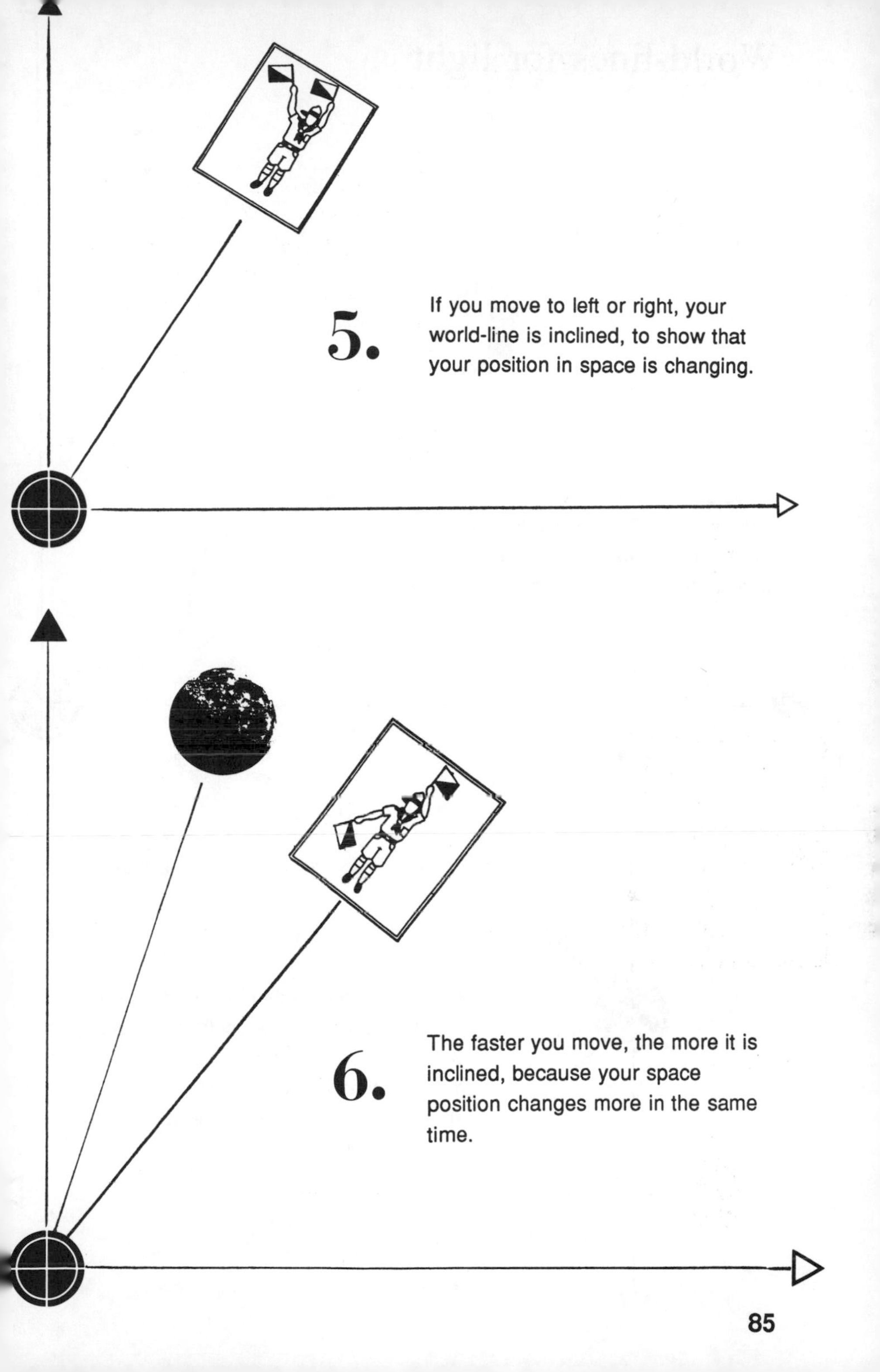

5. If you move to left or right, your world-line is inclined, to show that your position in space is changing.

6. The faster you move, the more it is inclined, because your space position changes more in the same time.

# World-lines for light

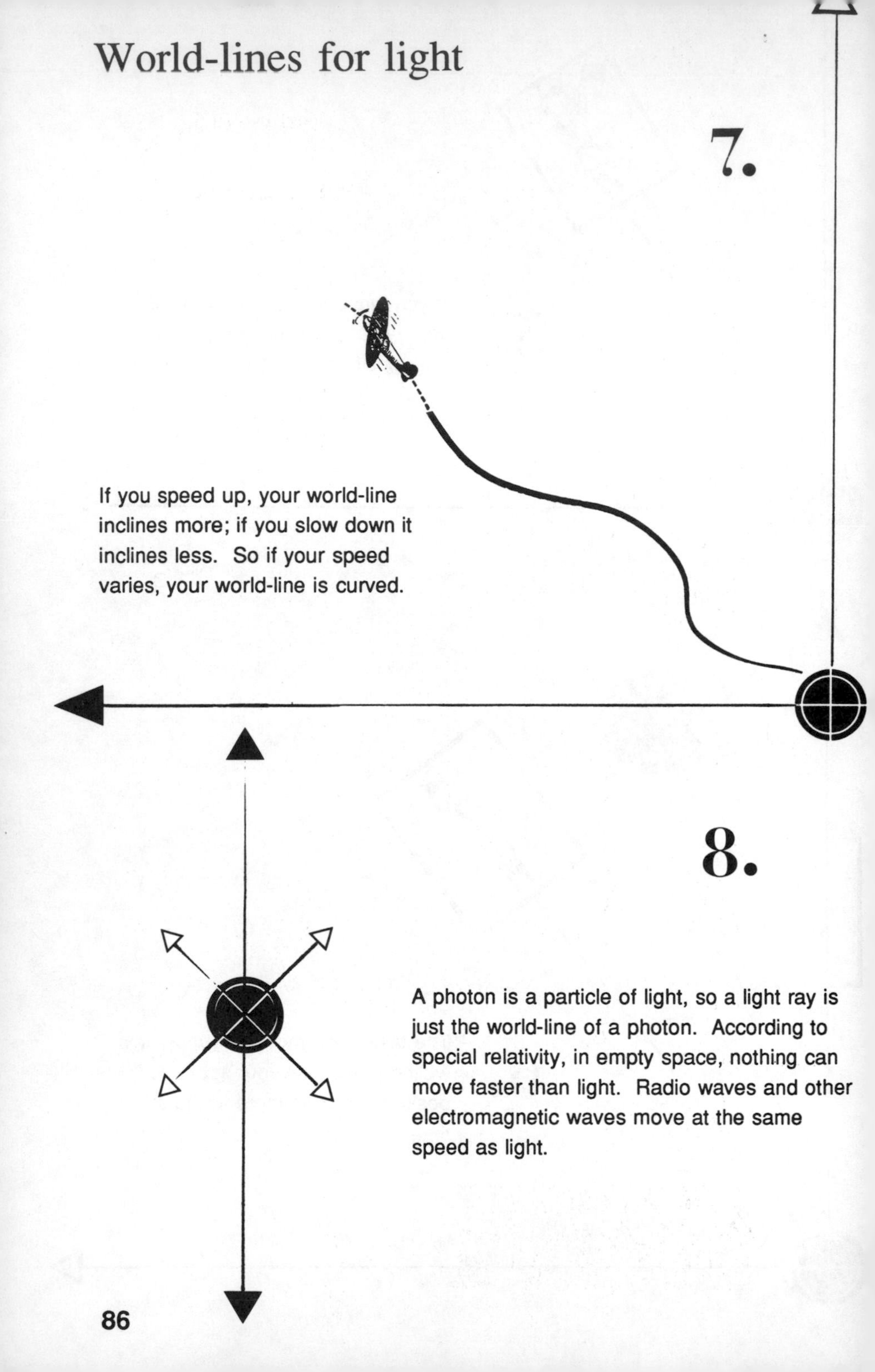


If you speed up, your world-line inclines more; if you slow down it inclines less. So if your speed varies, your world-line is curved.

A photon is a particle of light, so a light ray is just the world-line of a photon. According to special relativity, in empty space, nothing can move faster than light. Radio waves and other electromagnetic waves move at the same speed as light.

# Past and Future

## 9.

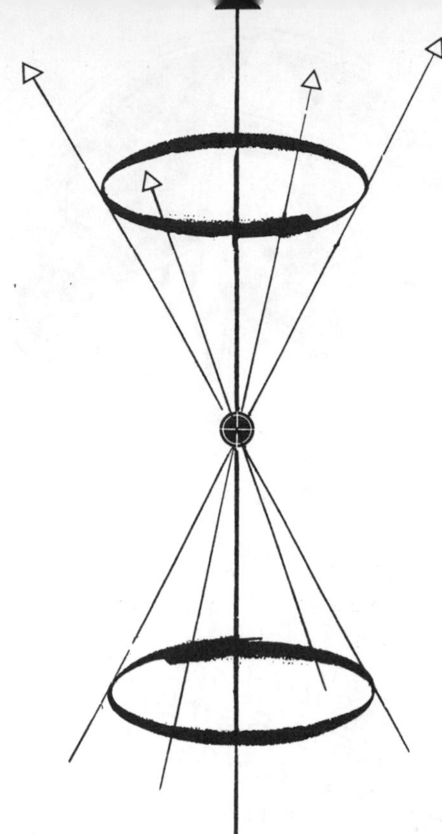

This space-time diagram, shows two space dimensions and the time, drawn in perspective. The light rays form a light cone, separated into two parts at the Here-and-Now point.

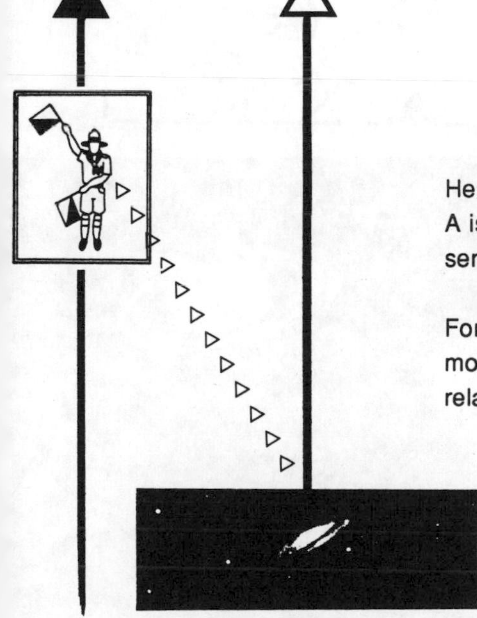

## 10.

Here are two world-lines, yours and A's. A is in the Andromeda galaxy. She sends you a radio signal.

For simplicity, you are both shown not moving. Actually, Andromeda does move relative to our galaxy.

## 11.

A is about 2,300,000 light-years away. You don't see her as she is now, but as she was, 2,300,000 years ago.

## 12.

If you send her a reply, it will take 2,300,000 years to reach her.

**Looking outward in space = looking backward in time.**

# 13.

The further away something is, the longer ago did it send out the light which you see.

# 14.

You can receive signals only from events within, or on, your past light cone.

# Newton's Law of Gravitation

The force of gravity keeps the planets moving in their paths around the Sun, and the satellites in their paths around the planets. This same force keeps the stars moving round in a galaxy, and the galaxies moving round in a cluster. These motions may be explained by Newton's law of universal gravitation.

Newton's law says that every object attracts every other object with a force proportional to their masses and inversely proportional to the square of the distance between them:

$$F = \frac{GMm}{r^2}$$

where ***F*** is the force, ***M*** and ***m*** the masses, and ***r*** the distance. The formula means that if you double either of the masses, it doubles the force; if you double the distance, it reduces the force to a quarter of what it was.

The law doesn't tell what the force is in any particular case. For that, you need to know the value of the constant G in front, which is called the constant of gravitation.

The value of G is approximately $G = 6.6726 \times 10^{-11}$ m$^3$/kg.sec$^2$

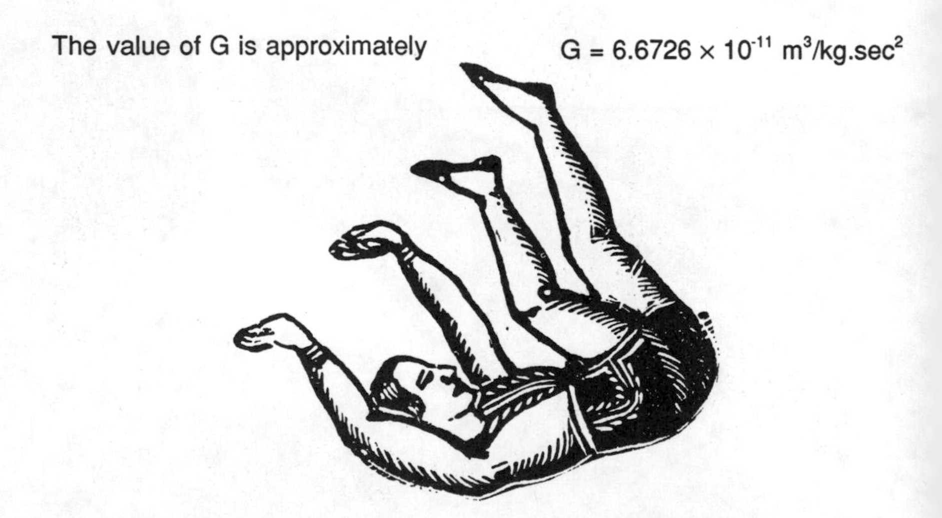

# How Attractive are you?

The gravitational attraction between two 400,000-ton supertankers sitting side by side in drydock is about the same as the weight of an adult.

The gravitational attraction between two consenting adults...

# Mass and Weight

Mass is not the same as weight.

Mass is a measure of an object's resistance to being accelerated

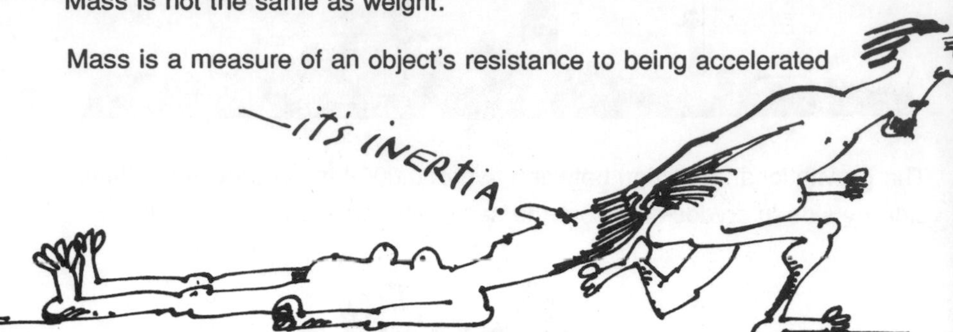


Mass enters Newton's equation:

**force = mass × acceleration.**

If here you put in the value of the acceleration due to gravity, this equation becomes:

**weight = mass × acceleration due to gravity.**

The same object will have different weights in different places, although its mass remains the same.

The mass of a particular astronaut will be the same on the Earth, on the Moon and in space.

The weight will be smaller on the Moon, where gravity is weaker than on the Earth,

Galileo showed by experiment that all bodies, or at least all he tested, fall equally fast, although he probably didn't do it by dropping weights from the top of the Leaning Tower of Pisa.

# All bodies fall equally fast

Suppose that M is the mass of the Earth, and m the mass which is dropped from the Leaning Tower. Then the law of gravitation says that

$$F = \frac{GMm}{r^2}$$

and the law of motion says that

$$F = ma$$

where ***a*** means the acceleration. If the ***m*** which comes into the law of gravitation is the same as the ***m*** which comes into the law of motion, then for motion under gravity

$$ma = \frac{GMm}{r^2}$$

and the m may be cancelled out of this equation, leaving

$$a = \frac{GM}{r^2}$$

The acceleration is the same, whatever the mass.

My theory allows this
but doesn't require it.

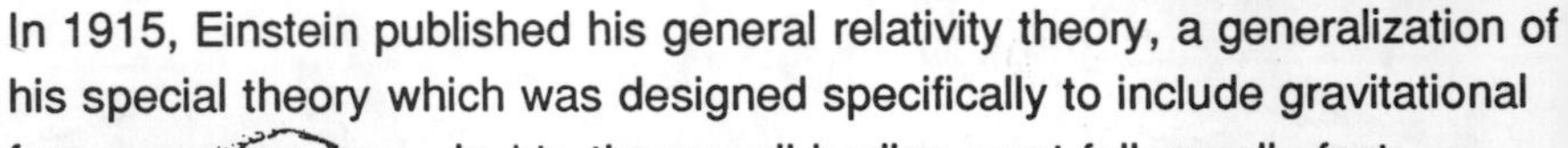

In 1915, Einstein published his general relativity theory, a generalization of his special theory which was designed specifically to include gravitational forces. In his theory, all bodies must fall equally fast. They can't help it.

# General relativity

General relativity is based on the principle of equivalence, which asserts that gravitational and inertial effects are indistinguishable.

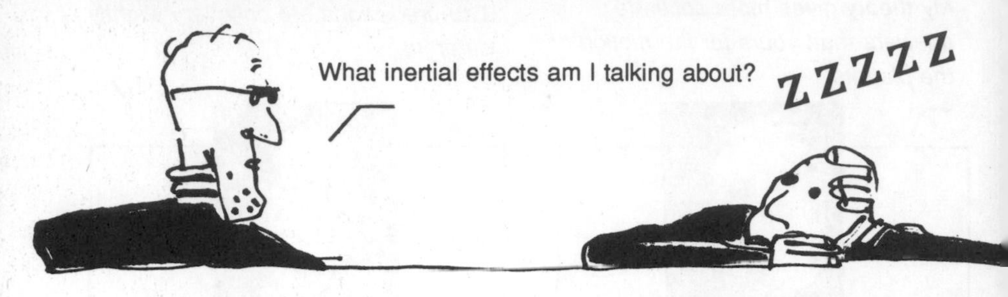


Acceleration, for example. Imagine yourself shut up in a space capsule. You can do experiments inside the capsule, but you can't see out. You feel that you're being pulled towards the floor of the capsule...

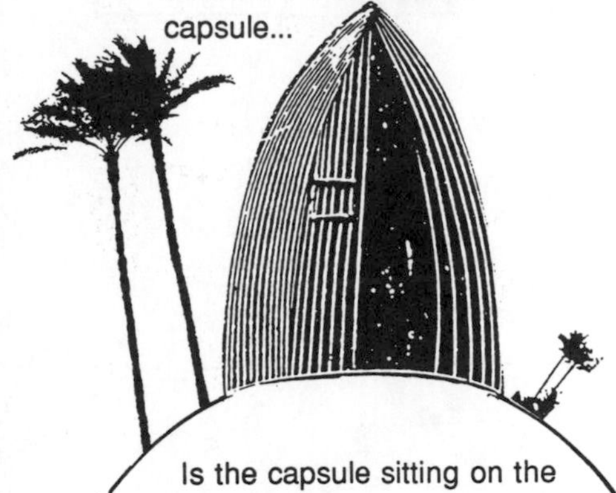

Is the capsule sitting on the Earth's surface, so that you feel the Earth's gravity...

or is it somewhere out in space, being boosted upwards by a rocket, so that you feel the acceleration?

Now suppose that there isn't any pull towards the floor.

Then the capsule must be floating out in space, with the rocket motors turned off.

Couldn't it be falling down an elevator shaft towards the centre of the Earth?

Air resistance would slow it down, and I'd still feel a pull.

Suppose the air had been evacuated from the elevator shaft. Then you wouldn't feel any pull.

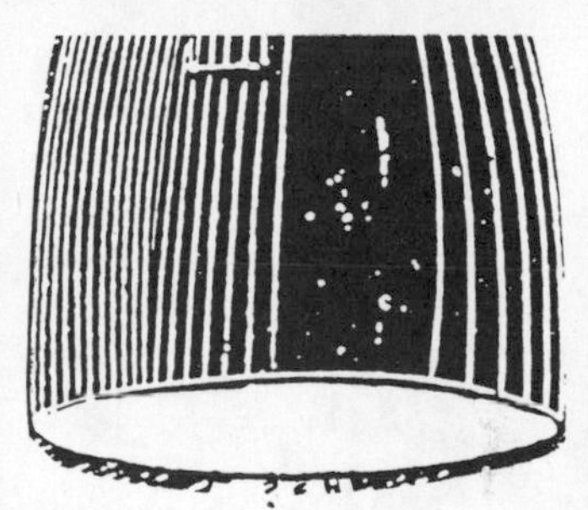

Then I couldn't tell whether the capsule was falling down the elevator shaft, or floating out in space.

**That's the principle of equivalence.**

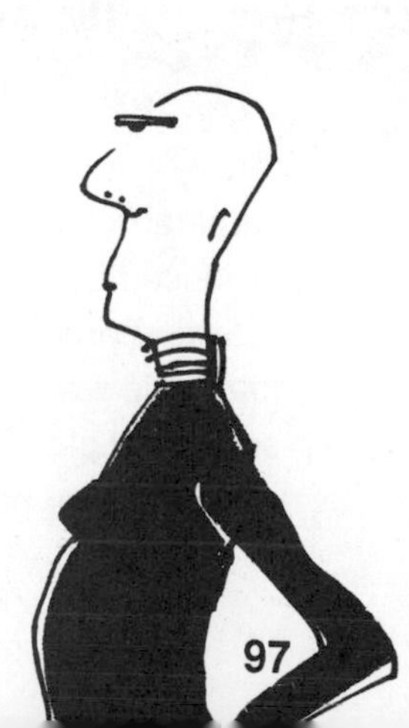

To accord with the principle of equivalence, general relativity describes gravitational effects and inertial effects in the same way, in terms of the "geometry" of space-time.

An ordinary map is flat. The space-time of general relativity is something like a map with the relief embossed on it. The geometrical shapes of the relief describe the gravitational field.

General relativity also predicts the existence of gravitational radiation, produced by moving masses. This is analogous to the electromagnetic radiation produced by moving charges in Maxwell's theory. Like electromagnetic waves, gravitational waves carry energy, and travel with the speed of light. The effects of gravitational waves have been observed, although they have not yet been detected directly.

Everything attracts everything else by gravitational forces. For example, matter attracts photons. To put it another way, gravitation distorts light-cones. The effect of this is that light rays passing near a massive object seem to bend. The bending of light rays by the Sun, predicted by Einstein, was first observed during an eclipse in 1919.

# Black holes

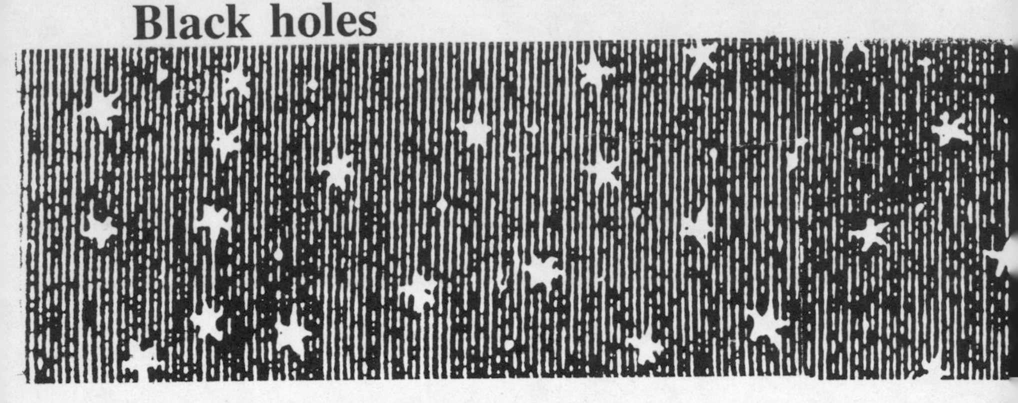

Stars attract the photons which they themselves have emitted.

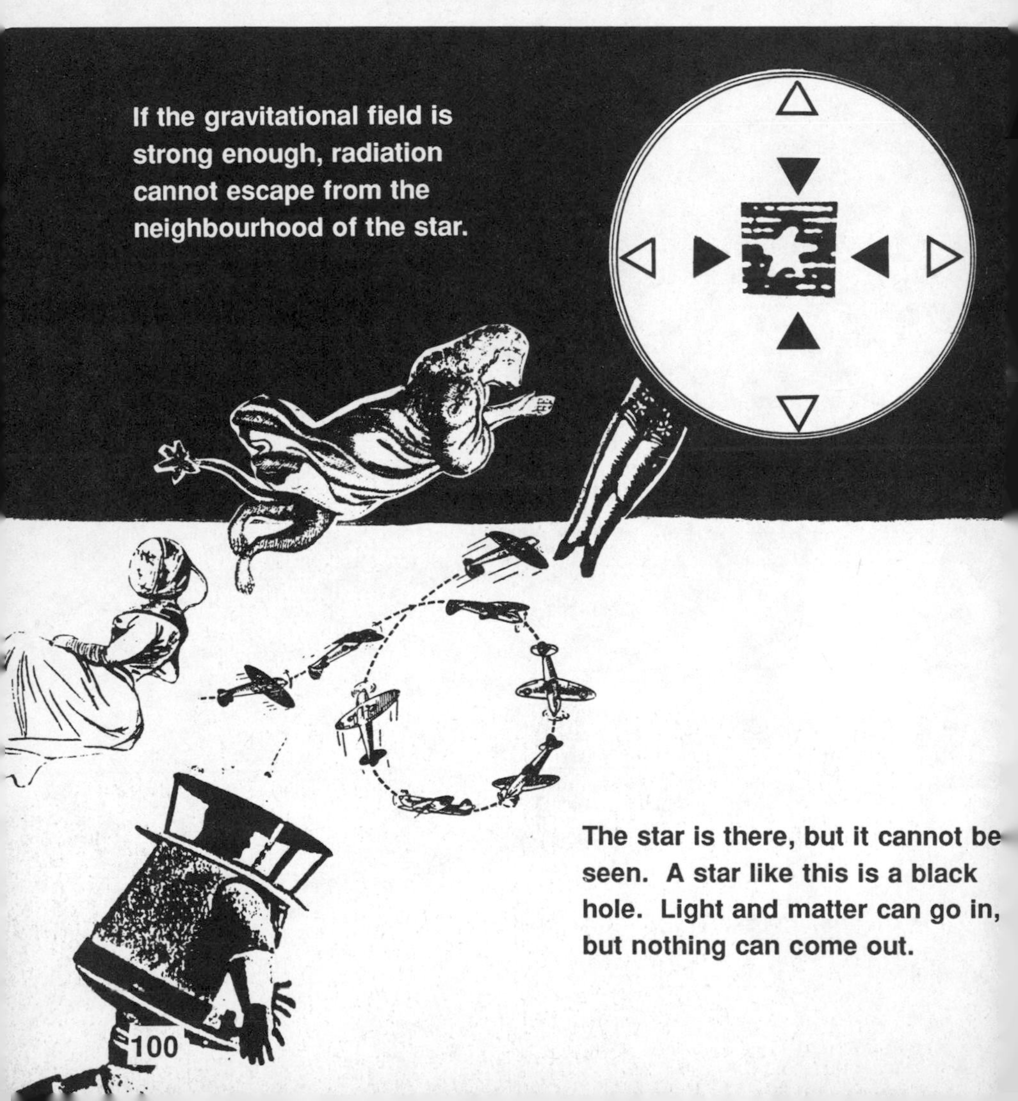

If the gravitational field is strong enough, radiation cannot escape from the neighbourhood of the star.

The star is there, but it cannot be seen. A star like this is a black hole. Light and matter can go in, but nothing can come out.

You can look at this in another way: If you throw a ball up into the air, gravity pulls it back.

How high it goes depends on how fast you throw it, and in which direction.

If you propel a satellite into space with a rocket, gravity pulls it back. Whether it comes back to Earth...

goes into orbit, or leaves the Earth's neighbourhood for good depends on how fast you propel it, and in which direction.

**The speed with which you need to propel an object straight up, in order that it can escape completely from the gravitational field, is called the escape velocity.**

At the surface of the Earth, the escape velocity is about 11 km per second. At the surface of the Sun, it is over 600 km per second.

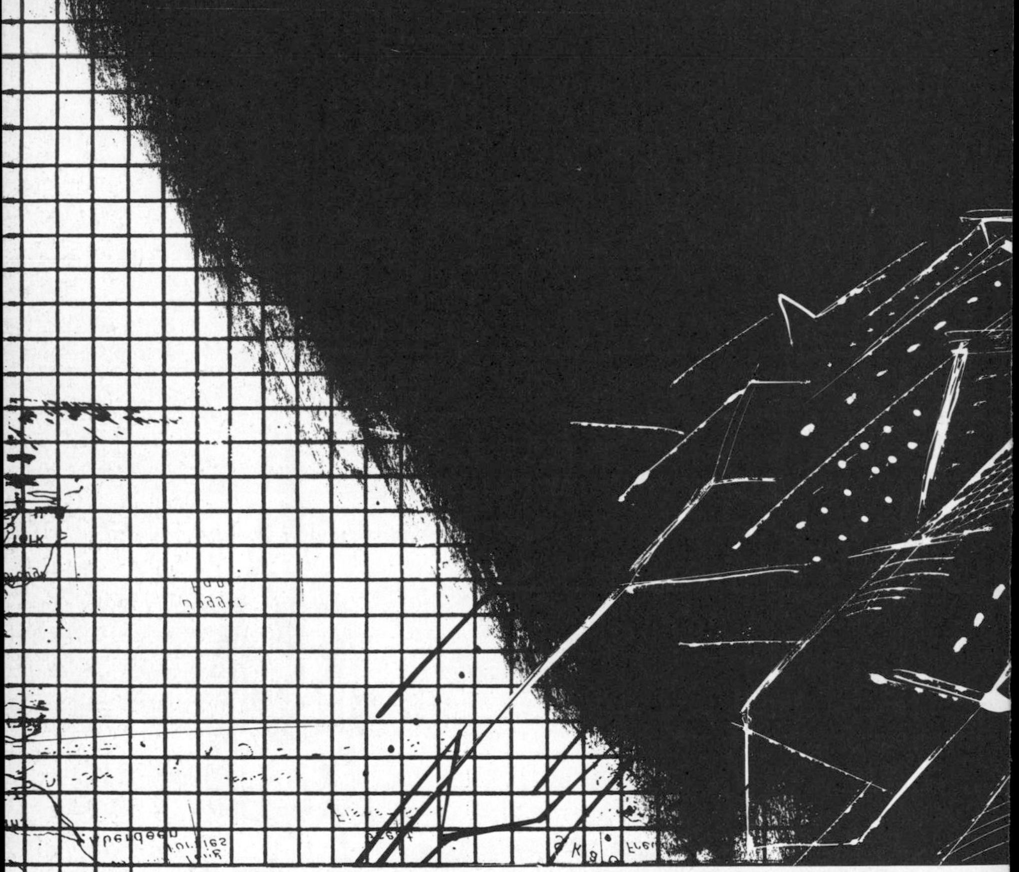

Some objects are thought to have collapsed to such a small size, for the[ir] masses, that the escape velocity is greater than the speed of light. Thes[e] are black holes. The gravitational field is so intense that nothing can escape from it. No particles. No radiation. Once in, you stay in. That's why you can't see a black hole.

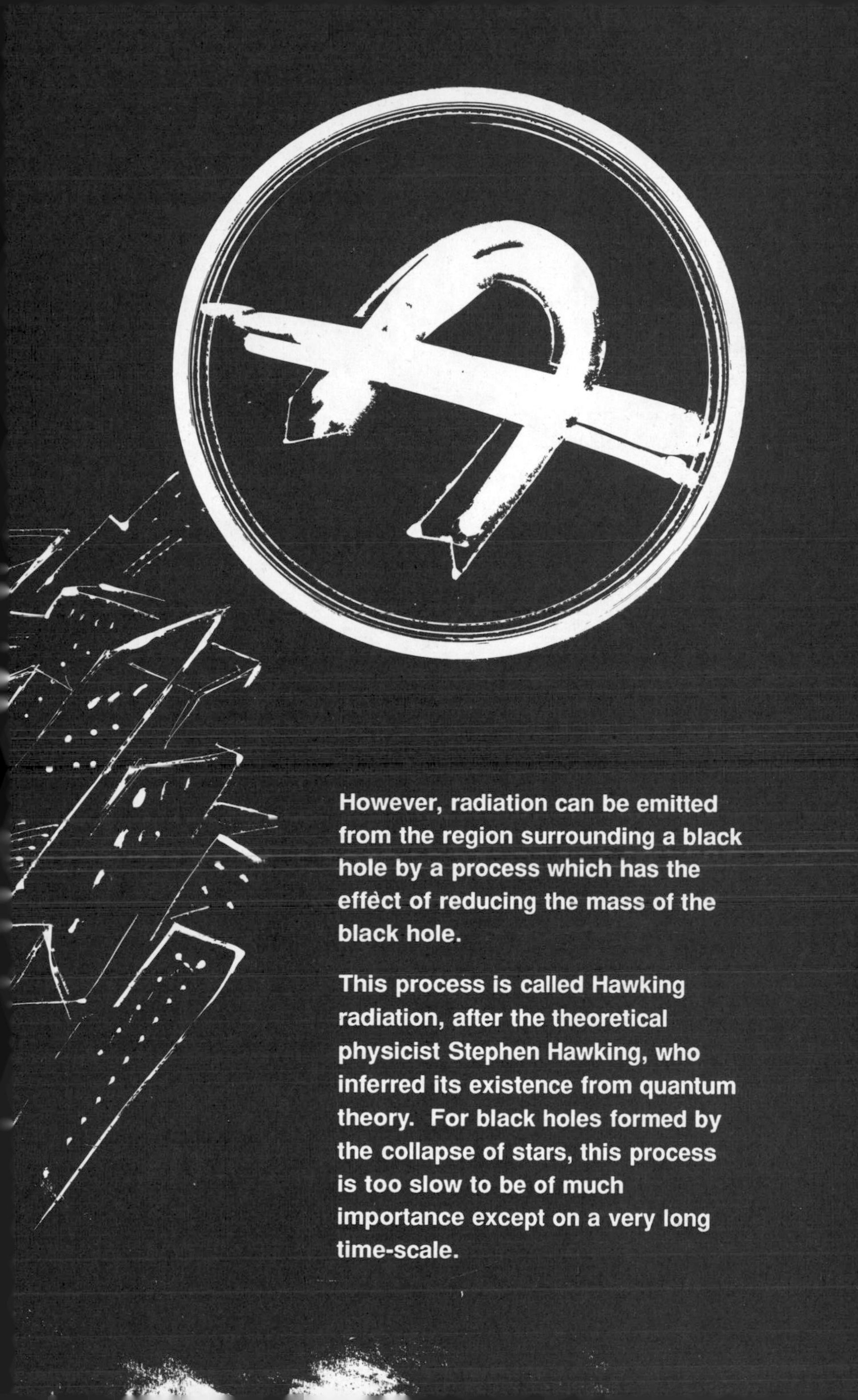

However, radiation can be emitted from the region surrounding a black hole by a process which has the effect of reducing the mass of the black hole.

This process is called Hawking radiation, after the theoretical physicist Stephen Hawking, who inferred its existence from quantum theory. For black holes formed by the collapse of stars, this process is too slow to be of much importance except on a very long time-scale.

## Stars and Black holes

Stars with large enough masses may end up as black holes, and there may be very massive black holes at the centres of some galaxies.

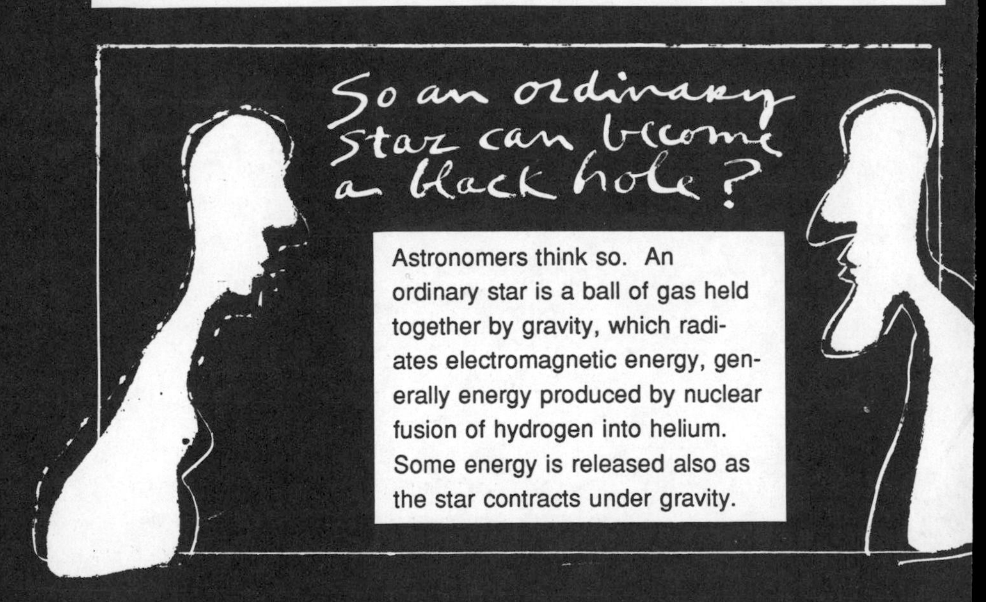


Astronomers think so. An ordinary star is a ball of gas held together by gravity, which radiates electromagnetic energy, generally energy produced by nuclear fusion of hydrogen into helium. Some energy is released also as the star contracts under gravity.

# Not making it.

"Brown dwarfs" with less than about 8% of the mass of the Sun do not achieve nuclear fusion.

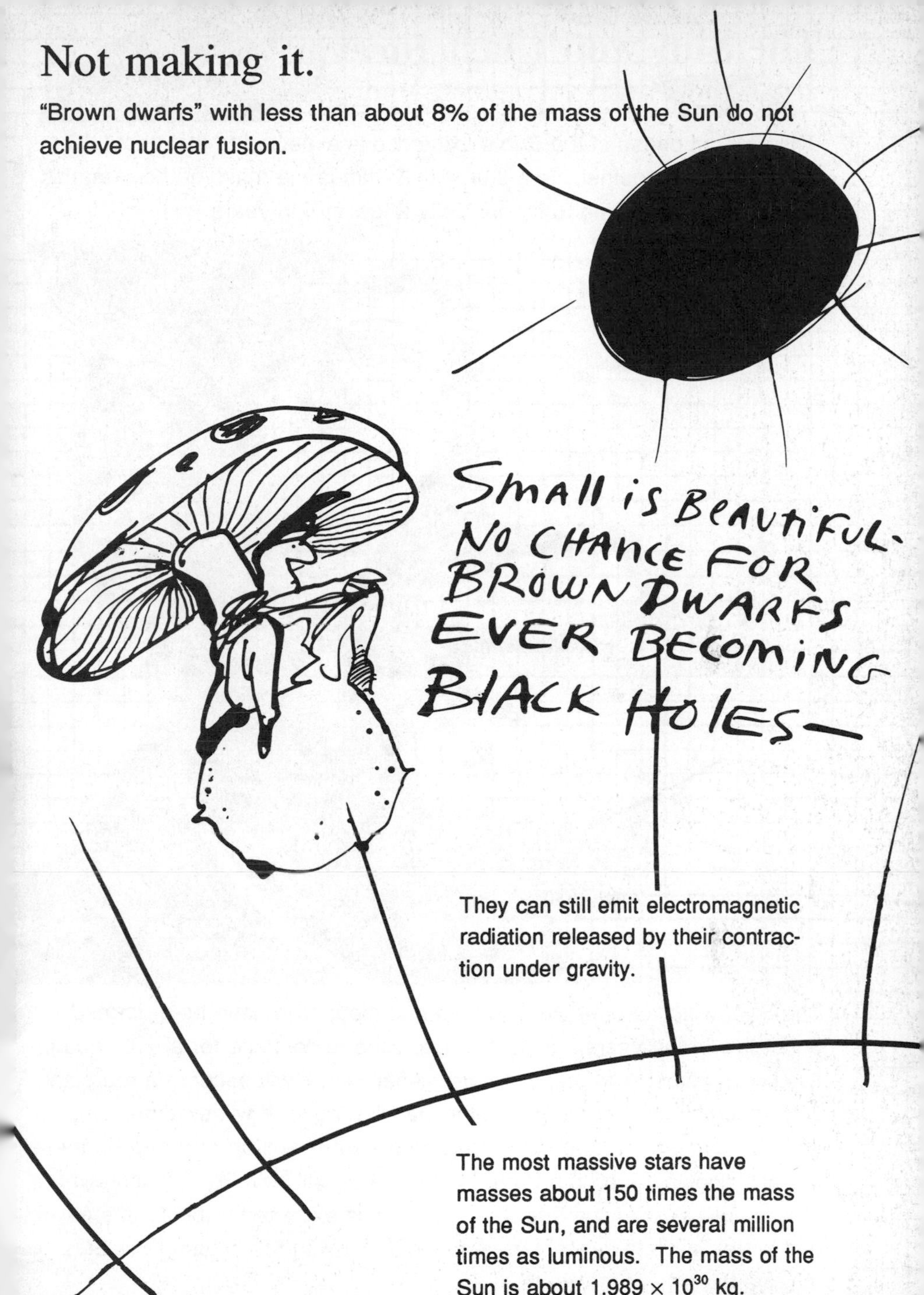


They can still emit electromagnetic radiation released by their contraction under gravity.

The most massive stars have masses about 150 times the mass of the Sun, and are several million times as luminous. The mass of the Sun is about $1.989 \times 10^{30}$ kg.

# The Sun won't last forever

The present phase of the Sun's existence is expected to last about 10 billion years altogether. In a star with 12 times the mass of the Sun, this phase lasts, according to theory, only a few million years.

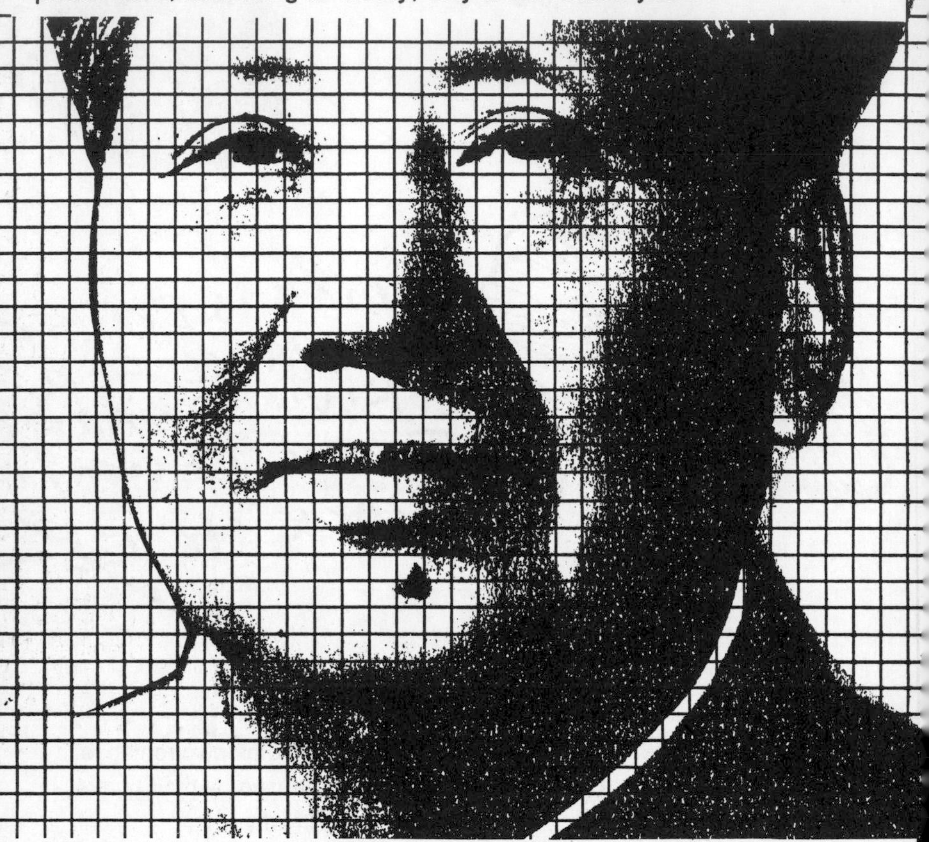

The Sun, which is a medium-size star, is thought to have been formed about five billion years ago, and is expected to continue for about another five billion years in its present state. After that, it will become a red giant star, much bigger and rather cooler than it is now. It will swallow up Mercury and Venus, and everything on the Earth will be burnt up. If there are any people left on the Earth by then, they will have to find somewhere else to live. Some time after that, the Sun is expected to become smaller and hotter, a white dwarf star. A white dwarf with the mass of the Sun may be the size of the Earth.

# Supernovae

The evolution of a star depends critically on its mass. The more massive the star, the more gravity compresses its material, the higher the central temperature becomes, and the more rapidly the nuclear reactions take place.

The subsequent history is also different. According to theory, a star with more than about 12 times the mass of the Sun eventually collapses under gravity, and then rebounds in an explosion, blowing off its outer layers and becoming very bright — a supernova. A supernova may become 10 billion times as bright as the Sun.

# Neutron stars

If what remains after a supernova explosion has a mass more than about 1.4 times and less than about three times the mass of the Sun, then according to theory the remains become a neutron star: an object composed entirely of neutrons. A neutron star with the mass of the Sun would have a radius of only 20 km. At the surface of a neutron star, the escape velocity is around half the speed of light.

If the mass which remains is greater than about three times the mass of the Sun, then it seems inevitable, from what is known now, that the remains of the supernova become a black hole. But no black hole has been definitely identified yet.

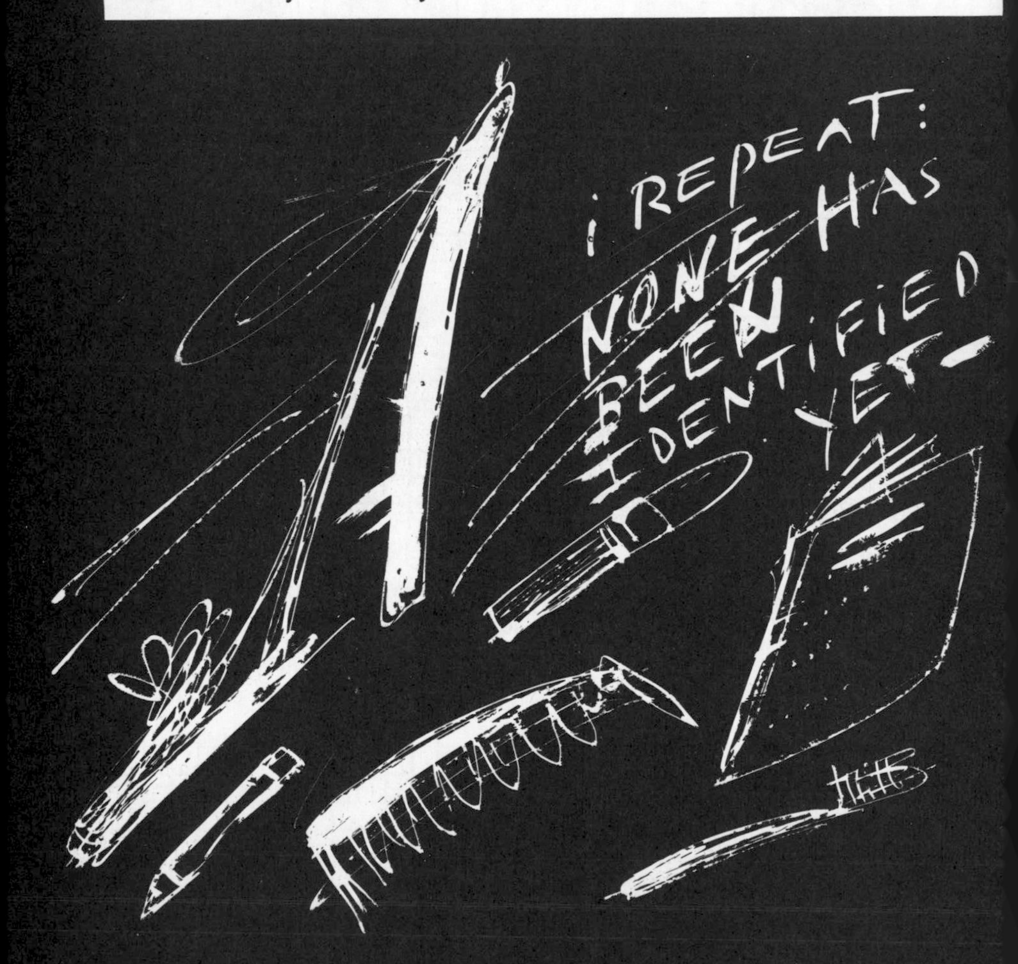

# HAS ANYBODY FOUND A NEUTRON STAR?

Jocelyn Bell, then a graduate student at Cambridge University, discovered a source of very regular radio pulses, one pulse every 1.3373 seconds.

THESE PULSES MAY COME FROM INTELLIGENT CIVILIZATION! SO WE'LL CALL THE SOURCE LGM1...

...FOR LITTLE GREEN MEN NO 1, OF COURSE.

But I was soon to discover three more pulsating sources, and about 450 more have since been found. Now they are called pulsars, and they are generally thought to be rotating neutron stars which send out narrow radio beams. Without any little green men to steer them.

Cambridge CLASS OF 67

# Quasars

The other favourite black hole candidates are quasars.

In 1963 this puzzle was solved:

*Maarten Schmidt*: This spectrum is just the hydrogen spectrum, with an enormous redshift.
These can't be ordinary stars. Let's call them quasi-stellar radio sources.

Or quasars, for short.

Today, most astronomers believe that quasars are very distant Active Galactic Nuclei (AGNs, for short). The most central one percent or so of a galaxy may be an AGN.

The light of a galaxy isn't just the combined light of its stars, if it has an AGN. An AGN doesn't emit normal starlight. It may produce as much radiation as a hundred billion stars. As much as a whole ordinary galaxy. Powerful radiation over a great range of wavelengths.

Many astronomers think that the quasars in AGNs are supermassive black holes, maybe several billion times the mass of the Sun. Gas from the surroundings is attracted by the black hole, which compresses and heats it. This hot dense material emits radiation very intensely.

There may be a black hole at the centre of our Galaxy, too.

It may be that some quasars are formed when two galaxies collide, and that small ones grow by sucking in gas from their surroundings. It seems that a few billion years ago, quasars were far more common than they are now. But none of this is guaranteed. Astronomers' ideas about black holes and quasars are very likely to change with time.

# Quantum theory.

It all started with black bodies. A black body is different from a black hole. A black body is an object which absorbs all the radiation which falls on it, reflecting nothing.

There are no perfect black bodies, but for example the inside of an electric oven which has been running steadily for some time with the door closed is a good approximation to a black body. A small region deep inside the Sun, where the gas and radiation are almost in equilibrium, would be another good approximation.

Black bodies absorb all the radiation which falls on them, but they emit radiation, too. According to theory, when a black body emits radiation, the radiation has a characteristic variation in intensity over the spectrum, and this intensity is completely fixed once you know the temperature of the body.

---

Max Planck devised the original quantum theory about 1900 to explain the spectrum of a black body.

The energy of my model radiating body can be emitted or absorbed only in definite packets, which I call quanta. The energy in each quantum is proportional to its frequency: **Energy = h x frequency**

h is called Planck's constant:

$h = 6.626176 \times 10^{-34}$ Joule seconds.

In 1913, Niels Bohr, by assuming that atoms exist only in definite quantum states, extended Planck's idea to explain the spectrum of hydrogen.

In the 1920s, Erwin Schrödinger, Werner Heisenberg and others devised a more elaborate and powerful theory, called

a theory generally applicable to atoms and their component parts and to radiation.

# Putting it all together

Special relativity and quantum mechanics can be unified, but there is no satisfactory synthesis of general relativity and quantum mechanics. Physicists have been trying to meld them together for over 50 years, without much success.

# The Planck time

The Planck time is the earliest time at which present-day physics makes any sense, the time at which the density of the contents of the Universe was the Planck density.

The Planck density $d_{Pl}$ and the Planck time $t_{Pl}$ are defined in terms of three fundamental constants of nature, Planck's constant h, the gravitational constant G, and the velocity of light c: $d_{Pl} = c^5/hG^2$ and $t_{Pl} = \sqrt{(Gh/c^5)}$. It's on the cover!

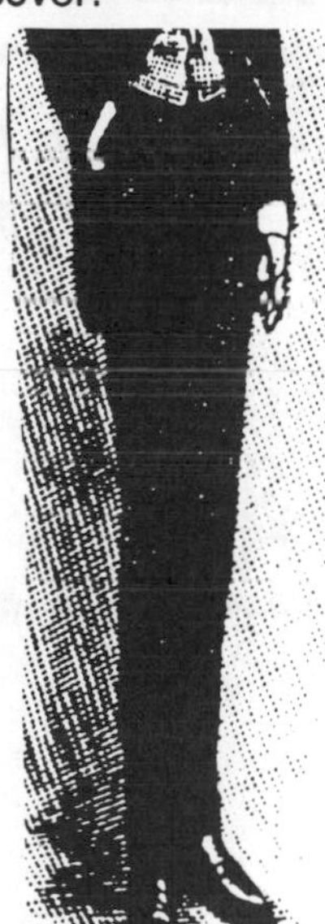

The Planck density is about $10^{93}$ times the density of water. Dense.

The Planck time is about $10^{-43}$ seconds. Short.

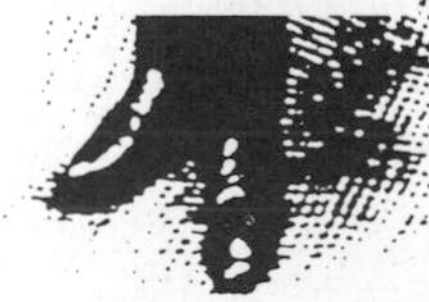

Present-day physics says nothing reliable about the nature of the Universe before the Planck time. Before the Planck time, the structure of space-time itself is uncertain.

*Chris Isham*: Our current understanding of space and time leaves much to be desired. There is no real evidence justifying the use of the Einstein equations [of general relativity] at subatomic scales.

This means that conditions for a beginning of the Universe cannot be specified theoretically. Cosmologists speculate about them anyhow.

Cosmologists make theoretical models — mathematical models — of the things they are trying to explain. Other scientists do the same.

The models are generally fairly simple. We incorporate what seem to be the most basic features of whatever is being modelled.

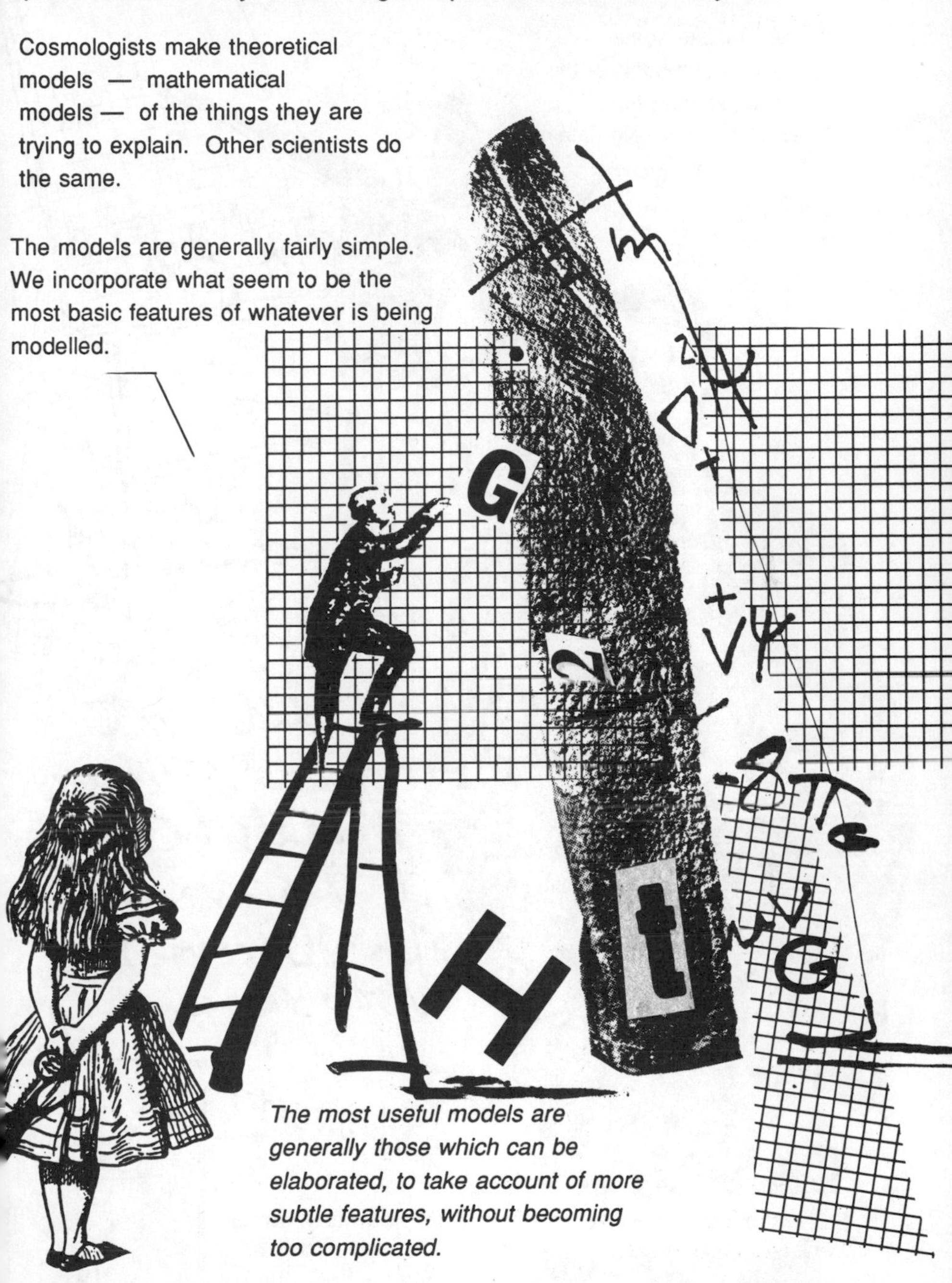


*The most useful models are generally those which can be elaborated, to take account of more subtle features, without becoming too complicated.*

# Friedmann models.

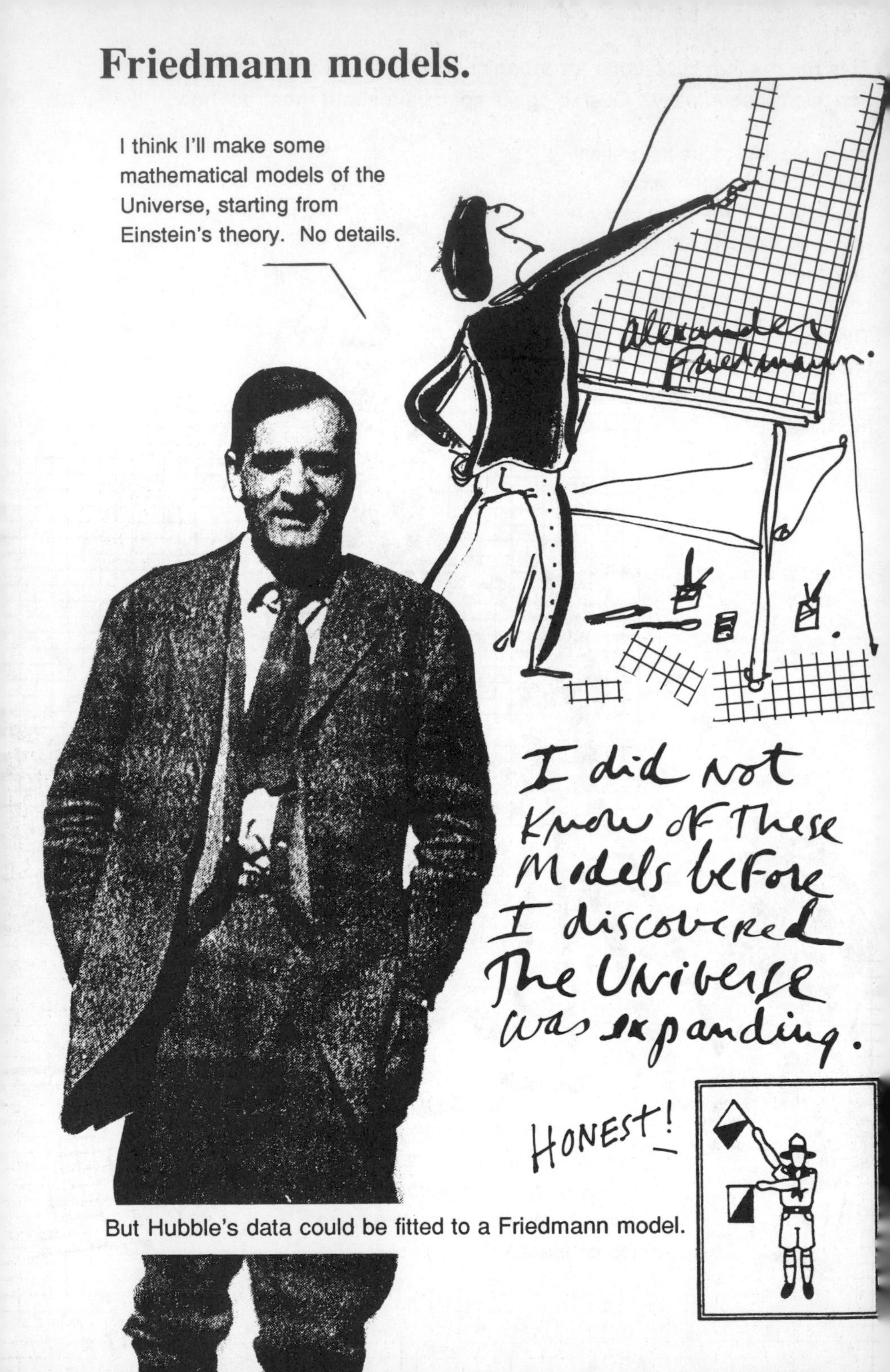


But Hubble's data could be fitted to a Friedmann model.

Campbells
CONDENSED
makes double
A FRIEDMAN MODEL IS LIKE THIN SOUP –
Models mustn't be stretched too far. Every model is limited in validity. The Friedmann models expand, and give Hubble's formula, but tell you nothing about the formation of the galaxies.
HERMANN BONDI
The Friedmann models do not present the detailed structure of the real Universe. A Friedmann model contains a completely uniform expanding (or contracting) distribution of structureless matter, without any stars or galaxies, let alone planets or people. The material of the Universe is there, but all the structure is totally smeared out.

# Friedmann models and

If you were in a Friedmann model, the Universe would look simplest to you if you were to go with the flow — move together with the smeared out material.

Imagine hypothetical astronomers in a Friedmann model, who go with the flow, without it making any difference to the model. Let's call them co-movers, because they move with the flow. You can work out what they would see, immersed in the model with all their instruments, but not having any effect on the way it behaves.

# Cosmic Time

The co-movers have a common time standard. And in most, but not all, Friedmann models there is an initial singularity from which the model can be said to begin.

Time measured from the initial singularity by the common standard is called cosmic time.

So in the mathematical model there may be a beginning of the Universe. This feature of the model cannot be used to picture the real Universe. A Friedmann model can have no validity before the Planck time.

The cosmological principle.

Each Friedmann model is exactly homogeneous.

This means that what co-movers see does not depend on where they are. At any chosen cosmic time, the model looks exactly the same to all co-movers.

Each Friedmann model is exactly isotropic.

This means that what co-movers see does not depend on which way they look. At any chosen cosmic time, the model looks exactly the same in every direction, to any co-mover.

The cosmological principle asserts that the Universe is homogeneous, and isotropic around every co-mover.

**According to the cosmological principle, every co-mover describes the history of the Universe in the same way, and makes no distinction between different directions in the model.**

The Friedmann models satisfy the cosmological principle exactly. The real Universe, on the other hand, satisfies the cosmological principle only approximately, for it is full of structure. It is clumpy at least up to the scale of superclusters of galaxies. Remember the Great Wall?

On the largest possible scale, the distribution of matter and motion seems to be rather accurately isotropic, but cosmologists differ about the degree of homogeneity.

**The largest scale is of the order of ten billion light-years, about as far as the most remote observable galaxies.**

HEllO, GRACE. YOU'VE NO DOUBT HEARD OF THE STANDARD COSMOLOGICAL MODEL?
YES JANE. IT'S JUST LIKE A FRIEDMANN, BUT WITH MORE STRUCTURE –
ON WASH DAY – OF All DAYS!

INDEED, GRACE. YOU ASSUME THAT THE EARLY UNIVERSE IS FIlLED WITH ElEMENTARY PARTICLES AND RADIATION IN A HOT, DENSE STATE...
...THAT ARE SUBJECT TO THE SAME LAWS OF NUCLEAR AND ELEMENTARY PARTICLE PHYSICS BASED ON QUANTUM MECHANICS...
·WHICH APPLY HERE ON EARTH

NEXT NIGHT
ISN'T THAT QUITE AN EXTRAPOLATION, PUMPKIN?
SURE, SUGAR. BUT IF WE DON'T MAKE THAT ASSUMPTION WE CAN'T SAY ANYTHING ABOUT WHAT HAPPENS.

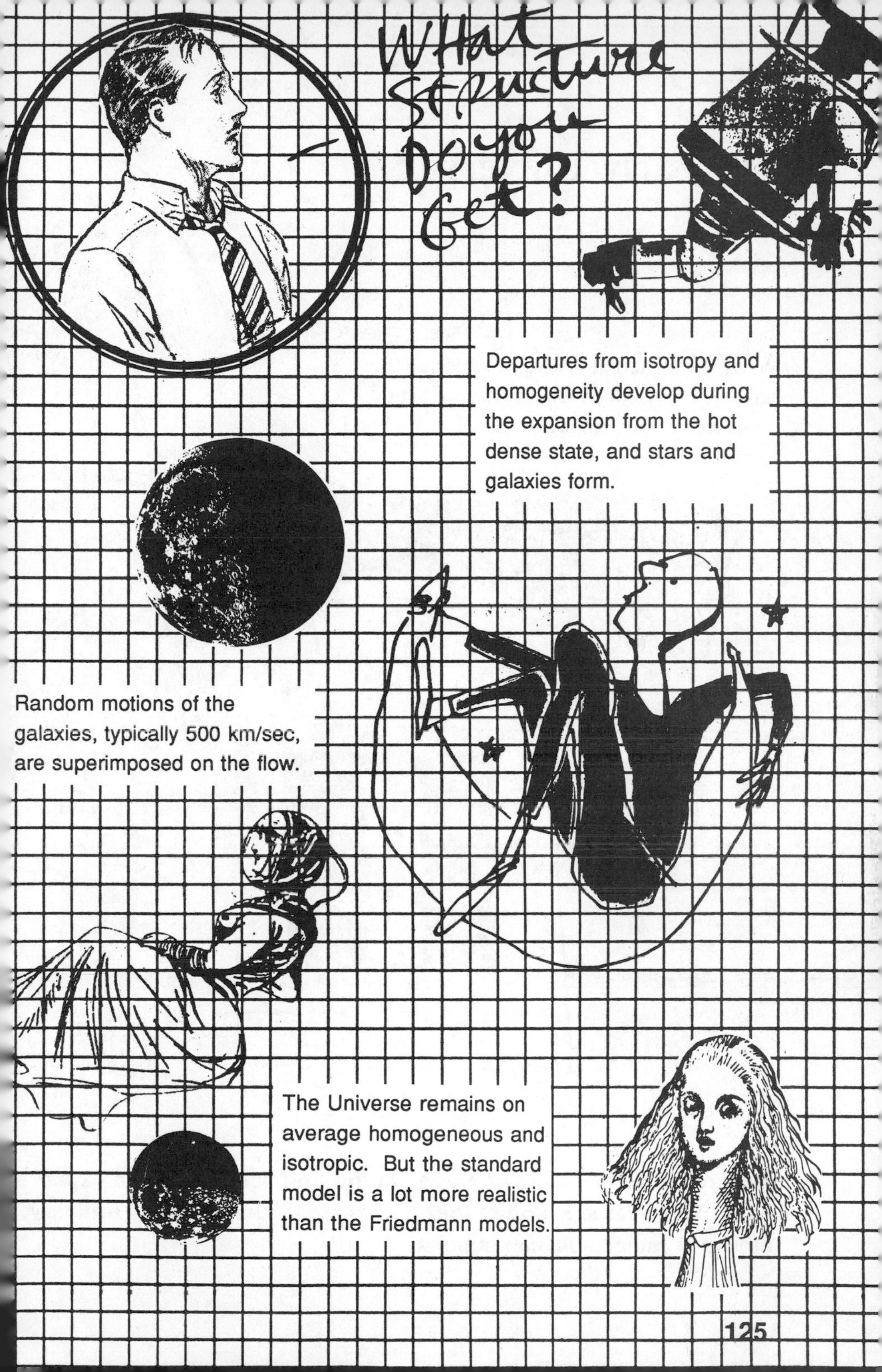


Departures from isotropy and homogeneity develop during the expansion from the hot dense state, and stars and galaxies form.

Random motions of the galaxies, typically 500 km/sec, are superimposed on the flow.

The Universe remains on average homogeneous and isotropic. But the standard model is a lot more realistic than the Friedmann models.

# The early Universe

Whatever started the expansion, perhaps an inflationary epoch, perhaps something more wild, is not intrinsic to the standard model.

**According to the standard model, in the first seconds of the expansion the Universe is so hot that no atomic nuclei, let alone ordinary atoms or molecules, can subsist.**

The standard model Universe cools as it expands. At a cosmic time of about one minute the synthesis of the lightest nuclei can begin. This process is called nucleosynthesis. Deuterium, helium and lithium are produced in the first thousand seconds.

As the expansion proceeds, the energy of radiation declines relatively and absolutely, and at a cosmic time of about 300,000 years, the formation of hydrogen atoms becomes possible.

matter

radiation

Subsequently the interaction between radiation and matter can to a great extent be neglected, and the remaining radiation travels on through the Universe practically without impediment. According to the standard model, it is this relic radiation which constitutes the cosmic background discovered by Penzias and Wilson. Because it was produced at a time when matter and radiation were practically in equilibrium, the relic radiation has a typical black body spectrum.

In the standard model, as the Universe continues to expand, galaxies and stars begin to form, largely under the influence of gravitational forces, and the Universe develops to its present state. The departures from homogeneity and isotropy become significant, but these properties are still, on a sufficiently large scale, preserved. The basic units of the Universe are galaxies.

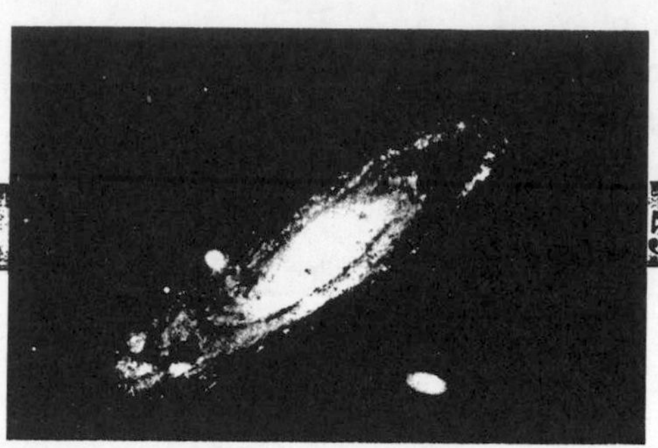

There are plenty of problems, too. For example, there is no single theory of the formation of galaxies and larger structures which fits all the observations.

Cosmologists cannot even agree whether galaxies formed first, and then collected into clusters, or whether clusters formed first, and then broke up into galaxies.

# The Future of the Universe

The standard model doesn't specify the amount of mass in a given volume — the density of matter. This has to be determined from observation. It makes a difference to what happens in the future.

Gravitational attraction between galaxies slows down the expansion. If the density is high enough, then gravitation is enough to halt the expansion altogether, and the Universe will after some time stop expanding and begin to collapse to a hot dense phase.

This final phase would not be just a big bang in reverse, because the collapsing material would be clumped, and there would be a lot of black holes; in its early stages the expanding Universe is smooth and unclumped.

The end point of this collapse has been called the big crunch. In the case where there is a big crunch, the Universe is called a closed Universe.

If the density is too low to halt the expansion, then the Universe will continue to expand forever. This is called an open Universe.

In the borderline case, in which the Universe just continues to expand forever, but at a rate which becomes smaller and smaller, the density is said to be at its critical value. This is called a flat Universe.

The present value of the average density, like the present value of Hubble's constant, is a matter of dispute. It is very probably somewhere between 2% and 100% of the critical value. Why it is uncertain you will see later.

---

# A big crunch

Suppose that the density of matter were so high that gravity could overcome the expansion. The Universe would continue to expand for some time from now, but at a decreasing rate.

**After the expansion stopped...**

As time went on the blue shift of each galaxy would increase. Ever more remote galaxies would appear blue-shifted.

As the Universe contracted, clusters of galaxies would merge into one another, and eventually galaxies themselves would merge. The present black body radiation, supplemented by the blue-shifted radiation from galaxies, would eventually become hotter and more intense than the radiation from stellar interiors. Everything would roast.

But that might not be the end. There might be a bounce: Some cosmologists suppose that the Universe oscillates. Instead of spreading out for ever from the big bang, or spreading out once from the big bang and then contracting once to a final big crunch, they conjecture that the Universe bounces back, each crunch turning into a new big bang. This is very speculative.

# The far future of an open Universe

Freeman Dyson once worked out what would happen in the far future, if the Universe were open. If you assume that

The laws of physics do not change with time.
The relevant laws are already known to us.
The Universe is open. Then...

**1**. Within $10^{14}$ years the longest-lived low mass stars will exhaust their hydrogen fuel and cool to very low temperatures. Stars of larger mass will reach their final states in shorter times.

**2**. After $10^{15}$ years, the Sun may be expected to have encountered another star closely enough to detach an Earth-like planet. The time until an encounter close enough for severe disruption of orbits will be considerably shorter.

**3**. After $10^{19}$ years, the central region of the Galaxy may be expected to collapse to a black hole, while stars in its outer regions are detached from it.

**4**. After $10^{20}$ years the Earth-Sun system would decay by radiation of gravitational waves. But the Earth will almost certainly escape from the Sun before gravitation pulls it in, unless the Sun escapes from the Galaxy with the Earth still in orbit.

**5**. After $10^{24}$ years stellar orbits around the Galaxy will decay by gravitational radiation.

**6**. After $10^{64}$ years a black hole of solar mass would decay by radiation. Decay of a black hole of galactic mass could take up to $10^{100}$ years. At the end of its life every black hole will become for a short time very bright. The cold expanding Universe will be illuminated by occasional fireworks for a very long time.

**7**. After $10^{1500}$ years all ordinary matter will have fissioned or fusioned to iron by radioactive processes.

What a way to go!

# How they see it

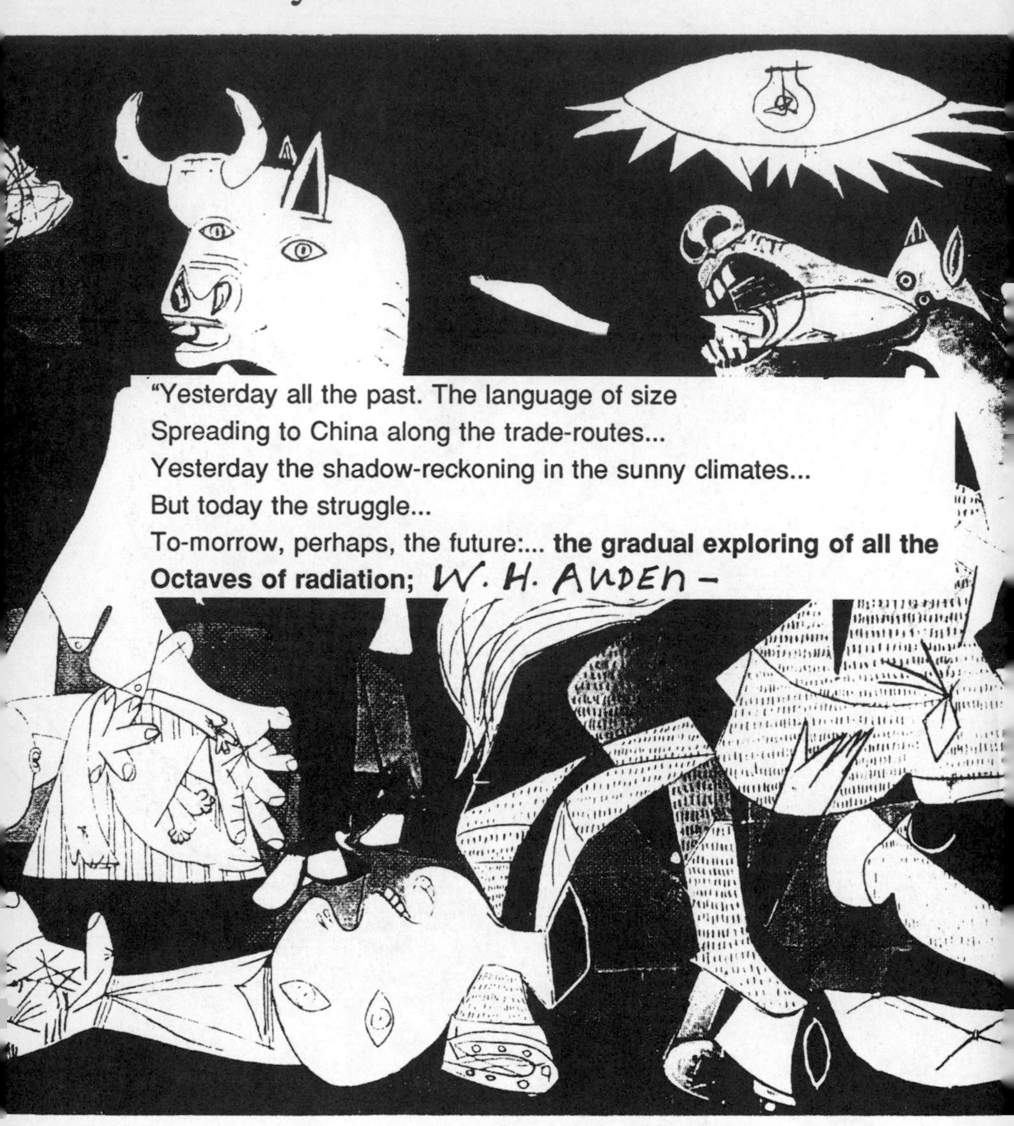


*Richard Goodman* (1939): the only poem by an Englishman anywhere near being a revolutionary poem.

*Auden* (1964): This is trash I am ashamed to have written.

How do they compare the standard model, and all that goes with it, with the actual Universe?

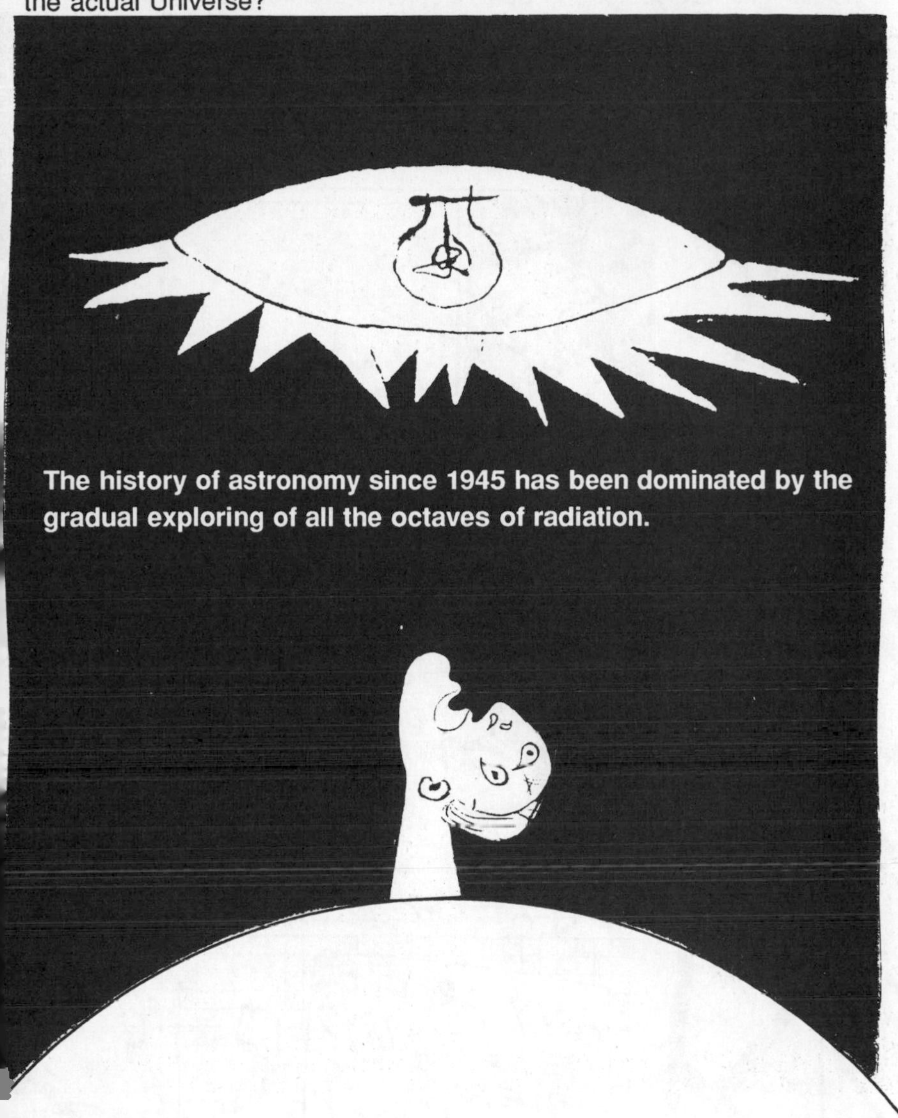


Radiation is absorbed by the Earth's atmosphere. The extent to which it is absorbed depends on the frequency. Visible light, and radio waves of frequencies between about 1.5 MHz and 300 GHz are impeded very little, but some other frequencies are almost completely absorbed before they reach ground level.

Until 1945 practically no astronomical observations were made except in visible light. Except for one Bell Telephone Laboratories researcher:

In 1933 Karl Jansky was searching for sources of interference with 20 MHz radio equipment being developed for ship-to-shore and transatlantic telephone communication.

The discovery made the front page of the New York Times.

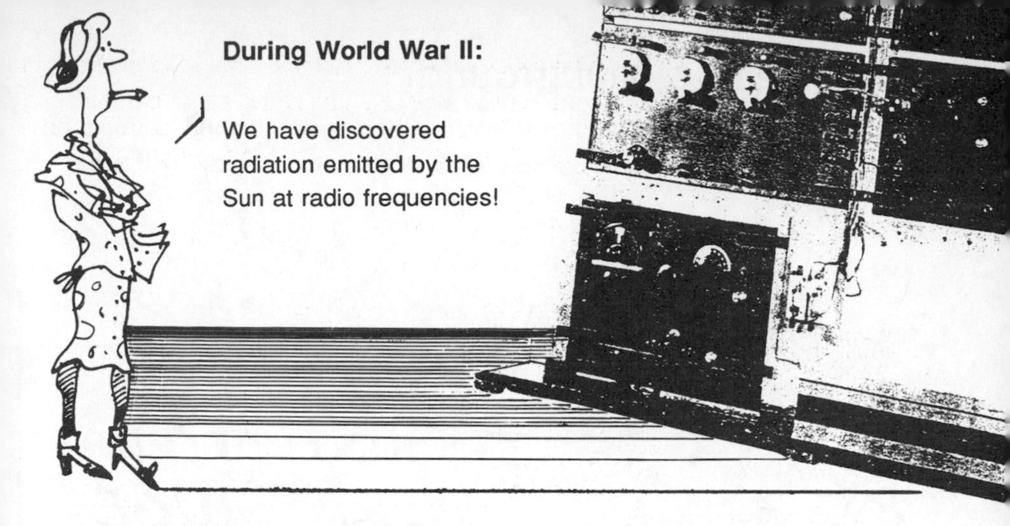


**After the War** there were both the scientists and the technology for a systematic exploration of the Universe at radio wavelengths.

In the 1970s it became possible to make observations from satellites. This removed all the limitations on astronomy which result from the absorption of incoming radiation by the Earth's atmosphere.

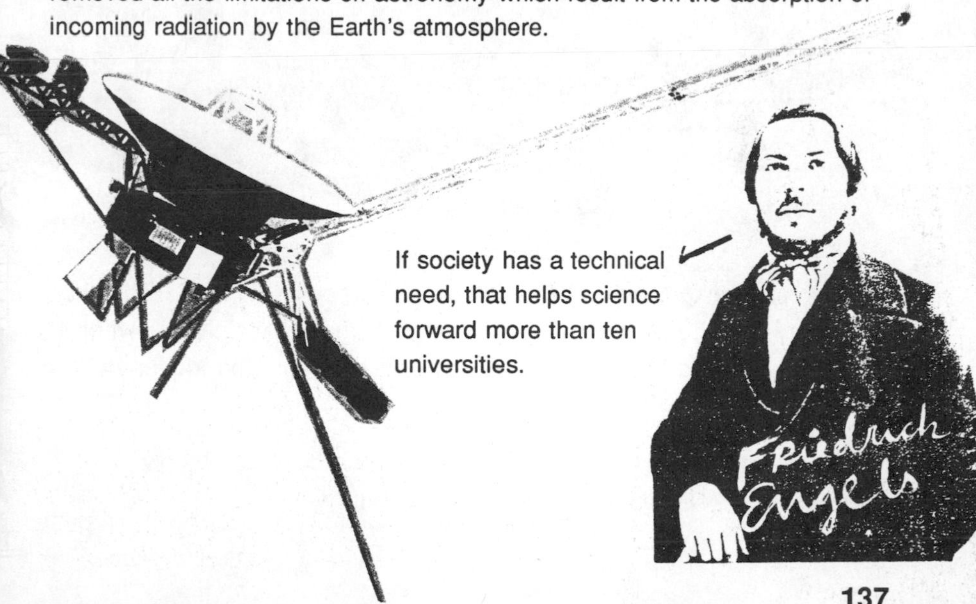

# The cosmic background.

The clearest evidence that the Universe is homogeneous and isotropic on a large scale is the cosmic background radiation, discovered by Penzias and Wilson.

The cosmic background radiation has a black body spectrum. This fits the theory that the radiation was produced in the early Universe when the contents were in a hot, dense condition and radiation and matter were in equilibrium.

Radiation with this kind of spectrum can be characterized by giving a single number, called its black body temperature.

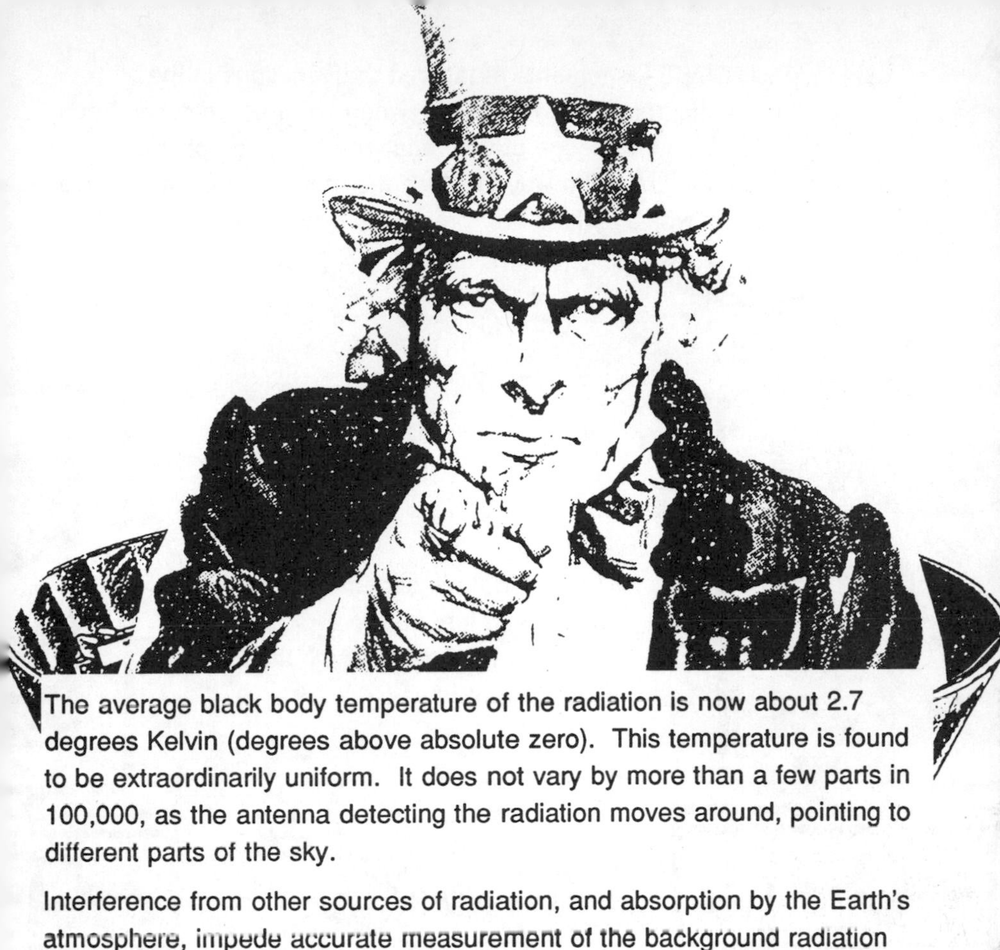

The average black body temperature of the radiation is now about 2.7 degrees Kelvin (degrees above absolute zero). This temperature is found to be extraordinarily uniform. It does not vary by more than a few parts in 100,000, as the antenna detecting the radiation moves around, pointing to different parts of the sky.

Interference from other sources of radiation, and absorption by the Earth's atmosphere, impede accurate measurement of the background radiation from the ground. Hence the interest for cosmology in the Cosmic Background Explorer (COBE) satellite, the first satellite dedicated to measurements of the cosmic background radiation. It was designed, built and in 1989 launched by scientists working for the US National Aeronautics and Space Administration (NASA).

Late in 1991, COBE instruments measured the variations in the temperature of the background radiation, when other sources had been eliminated. From a knowledge of the variations it may be possible to deduce something about the way in which the matter in the early Universe was clumped, of which nothing had previously been known.

The new data, if confirmed, should also lead to great advances in understanding the development of galaxies.

part or all of the variations found by COBE may come from gravitational waves...

not from clumping of matter in the early Universe.

The COBE results were surrounded with clouds of hype.

One cannot know anything of a past event unless some kind of signal could have been received from it.

Since no signal travels faster than the speed of light, there are events from which a signal cannot yet have arrived. You can know nothing about such events.

For example, you can know nothing about the Andromeda galaxy that happened less than about 2,300,000 years ago, because the Andromeda galaxy is about 2,300,000 light-years away.

There are distant events in well-separated parts of the sky which are so remote from one another that a signal cannot ever have been sent from one to the other.

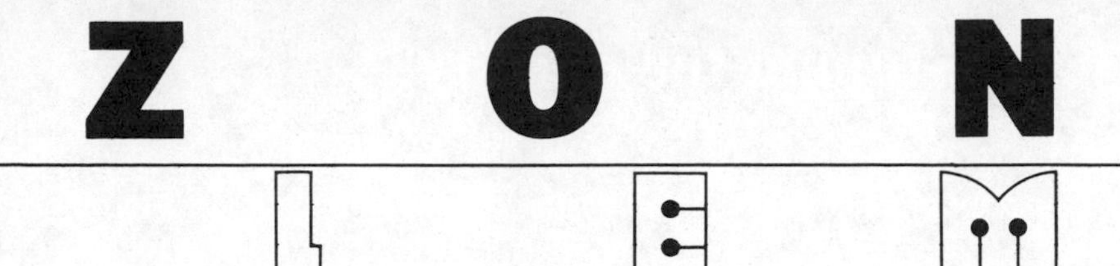


This raises a problem about the cosmic background radiation. Its black body temperature varies from point to point in the sky by only a few parts in 100,000.

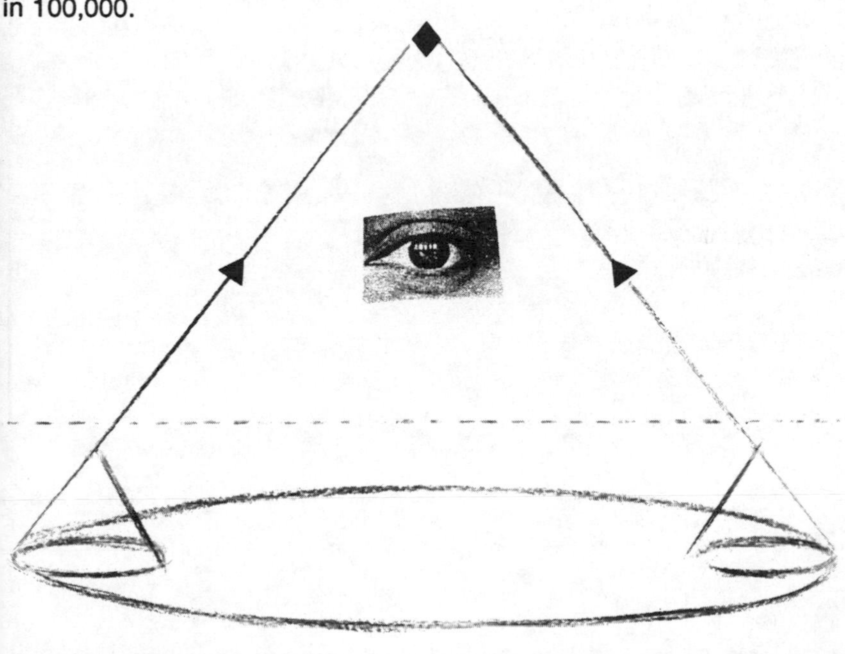

But the cosmic background radiation is thought to have originated only about 300,000 years after the big bang. At that time points in the sky which are well separated, as you see them now from the Earth, cannot have been within signalling range of each other, so that no physical interaction can ever have taken place between them.

There is no possible physical mechanism by which they could have arrived at equal temperatures. The initial uniformity of temperature has to be put into the standard model as a separate assumption. This is the horizon problem.

# Fine-tuning the density?

Another problem is presented by the density of matter.

The average density of matter in the Universe is not fixed by theory. The critical value, just low enough for the Universe to be able to go on expanding for ever, is an unstable one.

If the density ever differs slightly from critical, the difference increases with the passage of time.

From some knowledge of the present density, you can work the standard model backwards. It turns out that if the present value of the density is in the range considered probable, then at a cosmic time of one second it must have differed from critical by less than one part in $10^{15}$, and earlier it must have been very much closer still to the critical value.

This "fine tuning" of the average density is thought by many cosmologists to be inherently implausible.

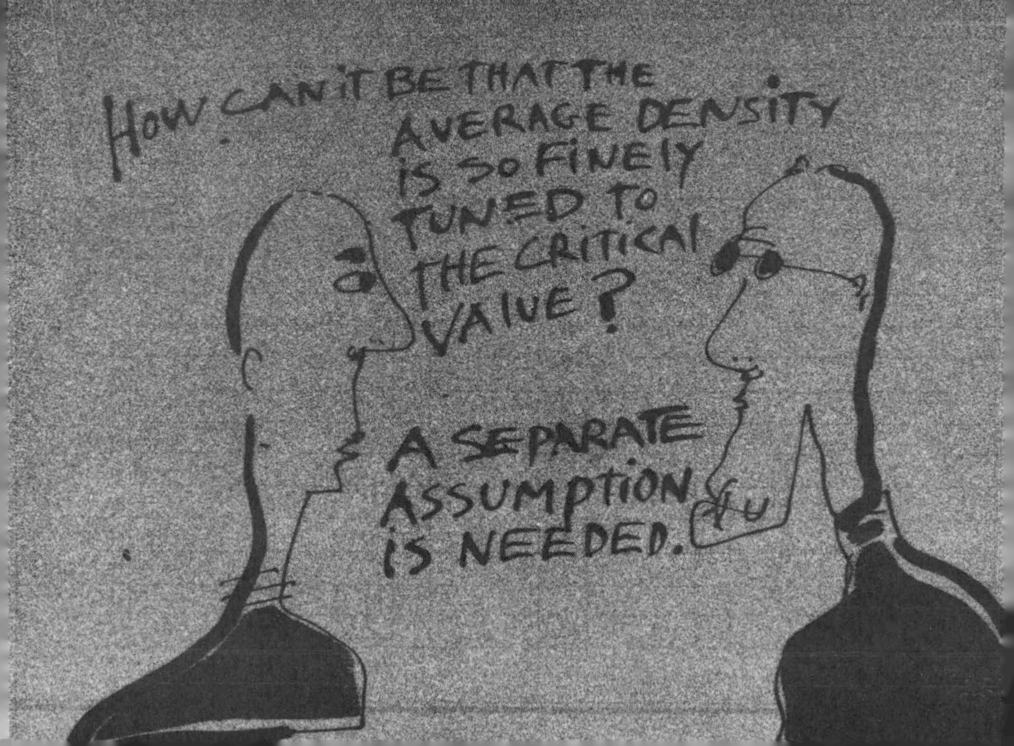

One way to deal with these problems would be to rely on developments before the Planck time, beyond the reach of present theory, to yield the needed isotropy and appropriate density, and to wait for a better theory.

Another way is to put them in by hand, as separate assumptions.

*David Layzer*: In the lexicon of science, the word Universe has no plural form. The initial conditions that define the Universe are no less unique than the laws. These cosmic initial conditions can't be deduced from anything else. They contain information about the world that we will never be able to explain.

On this view, the initial isotropy and homogeneity, and the initial value of the average density, are no more to be "explained" than are the equations which determine the subsequent evolution. They must necessarily be put in as separate assumptions.

# Inflation.

Many cosmologists reject these escape routes. They prefer a post-Planck-time solution to the problem, even if it involves conjecture. Such a solution is the inflationary cosmology, invented by A. Guth.

In 1980 Guth proposed conditions for the early Universe, based on the so-called Grand Unified Theories of elementary particles (GUTs).

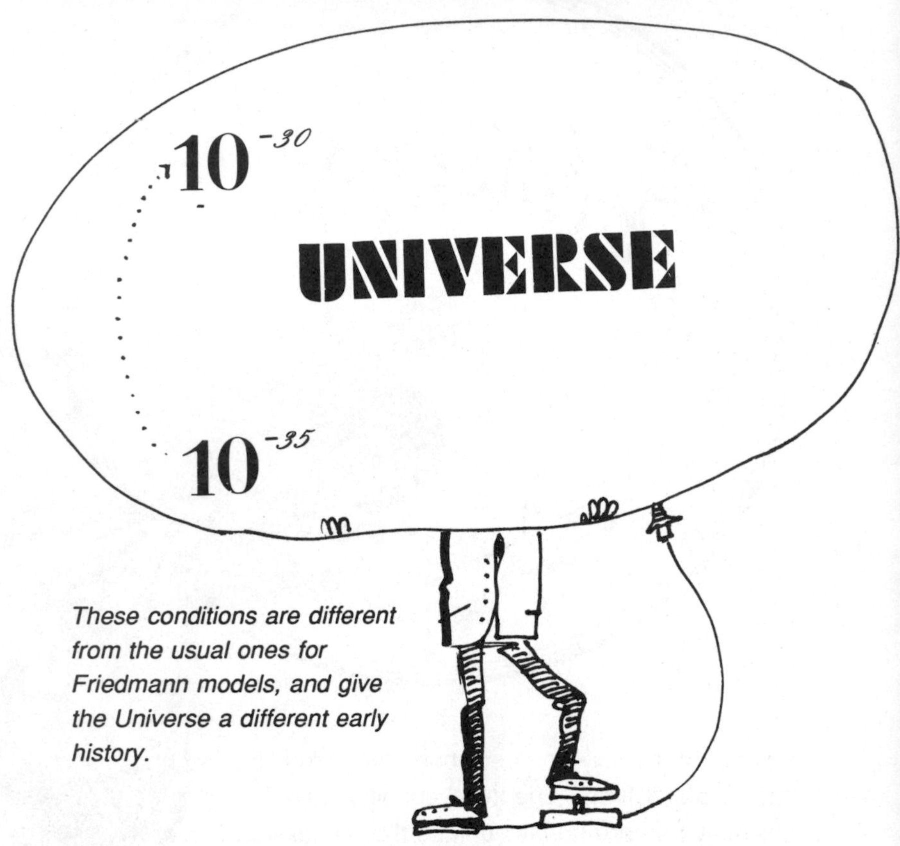


*These conditions are different from the usual ones for Friedmann models, and give the Universe a different early history.*

In the inflationary scenario an exceptionally rapid expansion of the Universe takes place from about $10^{-35}$ seconds to about $10^{-30}$ seconds after the big bang. This is the 'inflation'. Contrary to the usual case in the Friedmann models, the region which evolves into the presently observable Universe is such that all parts of it could have been in mutual communication for long enough to establish homogeneity and isotropy of the expanding material, which disposes of the horizon problem. Many variants on Guth's theory have been proposed subsequently.

In inflationary cosmology...

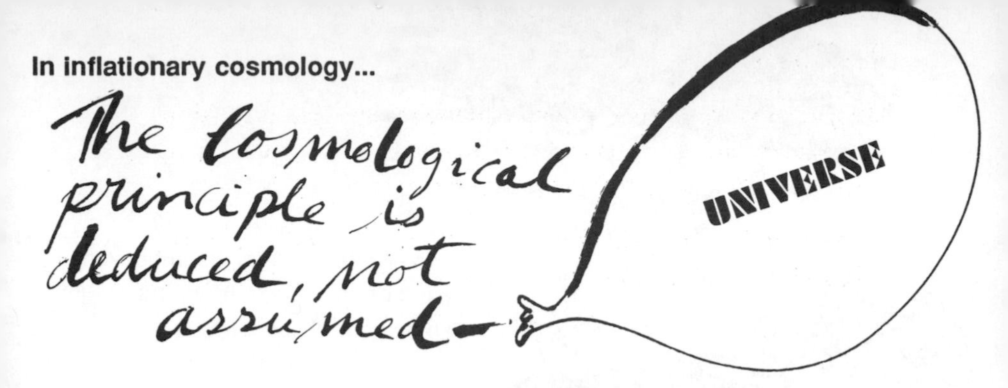


The isotropic property of the cosmic background radiation follows from the theory. The rapid expansion also forces the average density to be exactly critical.

The elementary particle theorists...

Here's an opening for us! We can work at far higher energies than are possible on Earth. The high temperatures in the early Universe are associated with such high particle energies.

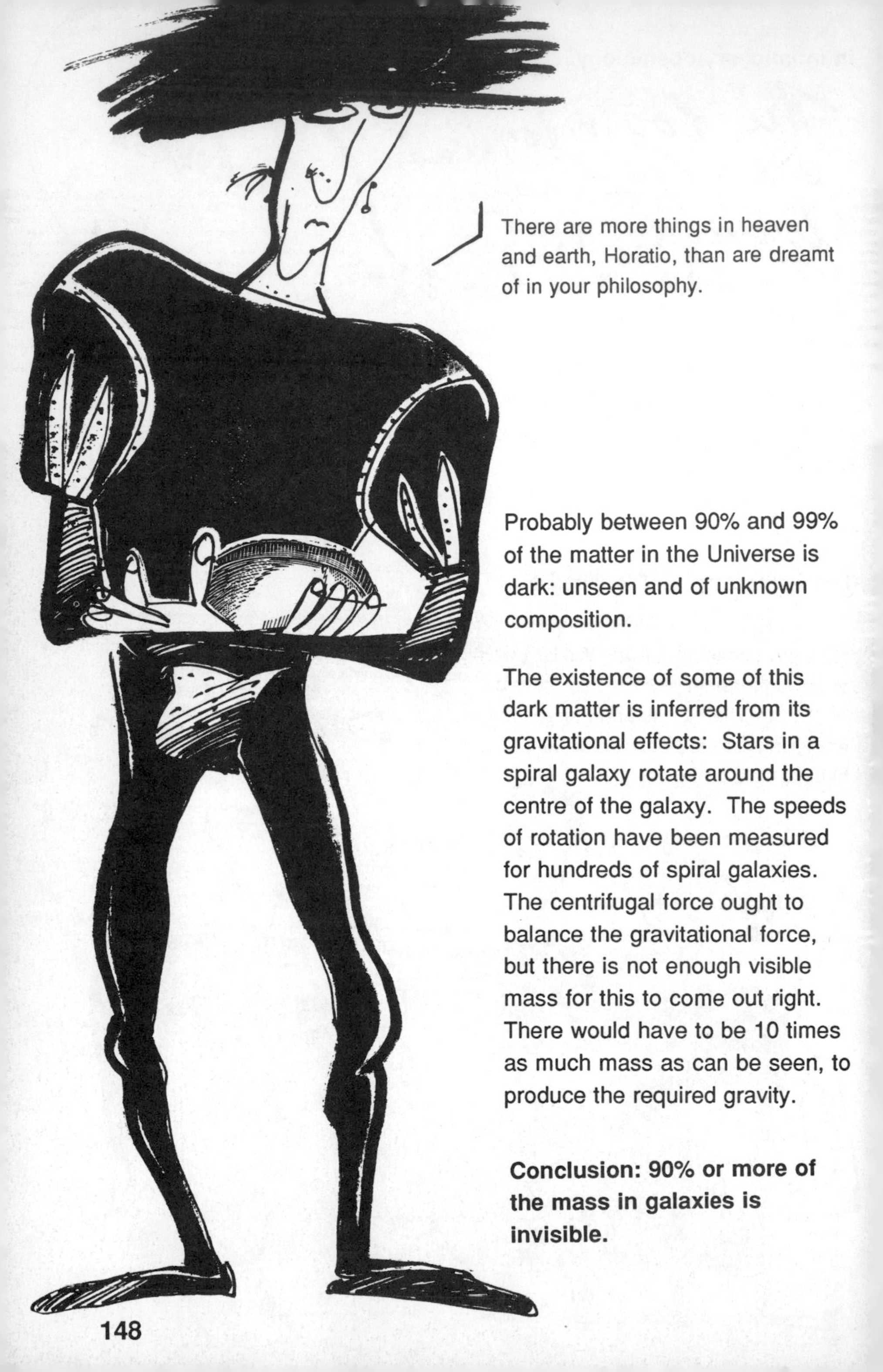


Probably between 90% and 99% of the matter in the Universe is dark: unseen and of unknown composition.

The existence of some of this dark matter is inferred from its gravitational effects: Stars in a spiral galaxy rotate around the centre of the galaxy. The speeds of rotation have been measured for hundreds of spiral galaxies. The centrifugal force ought to balance the gravitational force, but there is not enough visible mass for this to come out right. There would have to be 10 times as much mass as can be seen, to produce the required gravity.

**Conclusion: 90% or more of the mass in galaxies is invisible.**

Some of the invisible matter could be in brown dwarfs or smaller chunks of cool matter. Most of it must be further out than the stars in the galaxy. But the total mass and extent of an ordinary galaxy are not known with any precision. The same phenomenon is seen in clusters of galaxies.

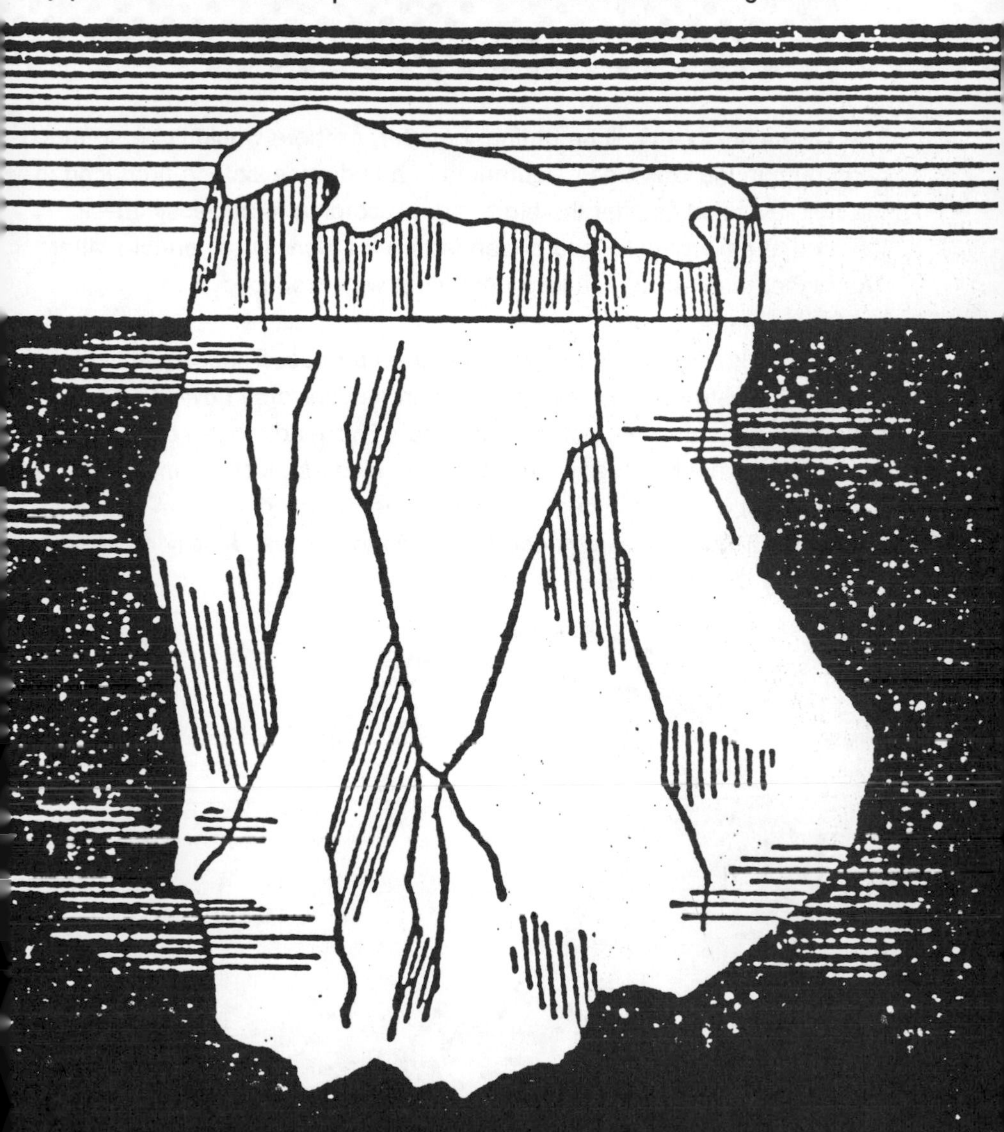

*Fritz Zwicky*: There must be far more matter in the Coma cluster of galaxies than can be seen, or the cluster would have blown apart by now.

The dark matter can't all be ordinary matter, or there wouldn't be so much deuterium in the Universe. Deuterium is a hydrogen isotope produced in the first few minutes after the big bang. According to nucleosynthesis theory, the amount produced depended on the density of normal matter. The higher the density, the less deuterium was produced.

As far as is known, deuterium has not been produced subsequently in stars, only destroyed. The amount of it in the Universe now can be estimated from measurements of spectra. This leads to the conclusion that there cannot be much more than ten times as much normal matter as is seen. But that isn't enough to explain the motion of clusters, and it's only about 10% of what is needed to reach the critical density required by inflationary theory.

The uniformity of the cosmic background radiation is further evidence for the existence of dark matter. It implies that matter was not very clumped together at the time the background radiation was produced, not clumped enough for there to have been time for the galaxies which we see to have developed out of the clumps.

Conclusion: Some of the matter in the Universe must be non-normal. If the inflation theory is right, more than 90% of the matter in the Universe is not normal matter.

# Hot and Cold Dark Matter

This dark matter may be hot or cold. Some people think it's a mixture. Perhaps 30% hot, 70% cold.

Hot and cold just mean fast and slow-moving. Cold dark matter, CDM, moves slowly, compared to light, relative to co-movers. Hot dark matter, HDM, moves fast. CDM is more susceptible to clumping by gravity than HDM.

HDM might be neutrinos, if they have any mass. Nobody's sure.

CDM might be exotic particles predicted by some form of GUTs. But none of these particles have been discovered yet.

Measurements of infra-red radiation from galaxies imply that it can't be only CDM.

***John Ellis*: Not only are we not at the centre of the Universe; we may not even be made of the same stuff as most of the Universe.**

# A difficulty for GUTs

GUTs predict that the proton is not a stable particle, but one which decays into other particles. According to GUTs, half the protons in any sample can be expected to decay into lighter particles in about $10^{32}$ years.

The processes which GUTs predict have not yet been observed. If GUTs go, cosmological theory is in trouble.

# **BIG** numbers — **BIG** coincidences

There are some big ratios in cosmology. For example, the ratio

$$\frac{\text{time since the big bang}}{\text{time for light to cross a hydrogen atom}}$$

is between $9\times10^{35}$ and $18\times10^{35}$, depending on how long it is since the big bang. And

$$\frac{\text{electric repulsion between two protons}}{\text{gravitational attraction between them}}$$

is about $12.4\times10^{35}$.

If you use electrons instead of protons, the second ratio would be about $4.18\times10^{42}$, over three million times bigger. How close the two numbers come depends on what exactly you choose to compare. But conventional theory offers no reason why they should be at all similar.

Here's another big one:
The ratio

$$\frac{\textit{Mass of the observable Universe}}{\textit{mass of the proton}}$$

depends on what the average mass density in the Universe is, and on what Hubble's constant is, but it's probably within a power of ten or so of $10^{79}$. Not too far from the square of either of the others.

---

It should be possible to calculate these big numbers from physical theory. I presume that all the numbers, except the age of the Universe, are constants. I have worked out a theory to predict these constants.

**Arthur Eddington (1936): The number of protons in the Universe is exactly**

And there are the same number of electrons.

**Arthur Eddington (1944): Sorry. I left out a factor ¾. The number of protons in the Universe is actually**

And as I said, there are the same number of electrons.

Few physicists found Eddington's arguments persuasive, partly because he had sometimes to adjust his results to fit fresh data.

*Paul Dirac* (1938): An accidental correspondence between very large numbers seems unlikely. I suggest that these large numbers are not constants, as Eddington proposes, but increase as time passes. There is some unknown connection between the quantities involved which has the consequence that the relations between the large numbers always remain valid. For this to be so, the gravitational force must gradually get weaker, compared to the electromagnetic force.

*Bob Dicke* (1961): An accidental correspondence between these numbers would indeed seem very unlikely if there were no reason for any of them to have one value rather than another. But there is a reason: The age of the Universe now can't be too small or too big.

It has to be old enough for elements other than hydrogen to exist — it is well known that carbon is required to make physicists. And it has to be young enough for there still to be a hospitable home for us in the form of a planet going round a luminous star. So the time since the big bang cannot be too small or too big. It has to be comparable to the lifetime of a Sun-like star.

# Anthropics

*Brandon Carter* (1974): Dicke's argument is an illustration of what I call The Weak Anthropic Principle (WAP): *Our location in the Universe is necessarily privileged to the extent of being compatible with our existence as observers.*

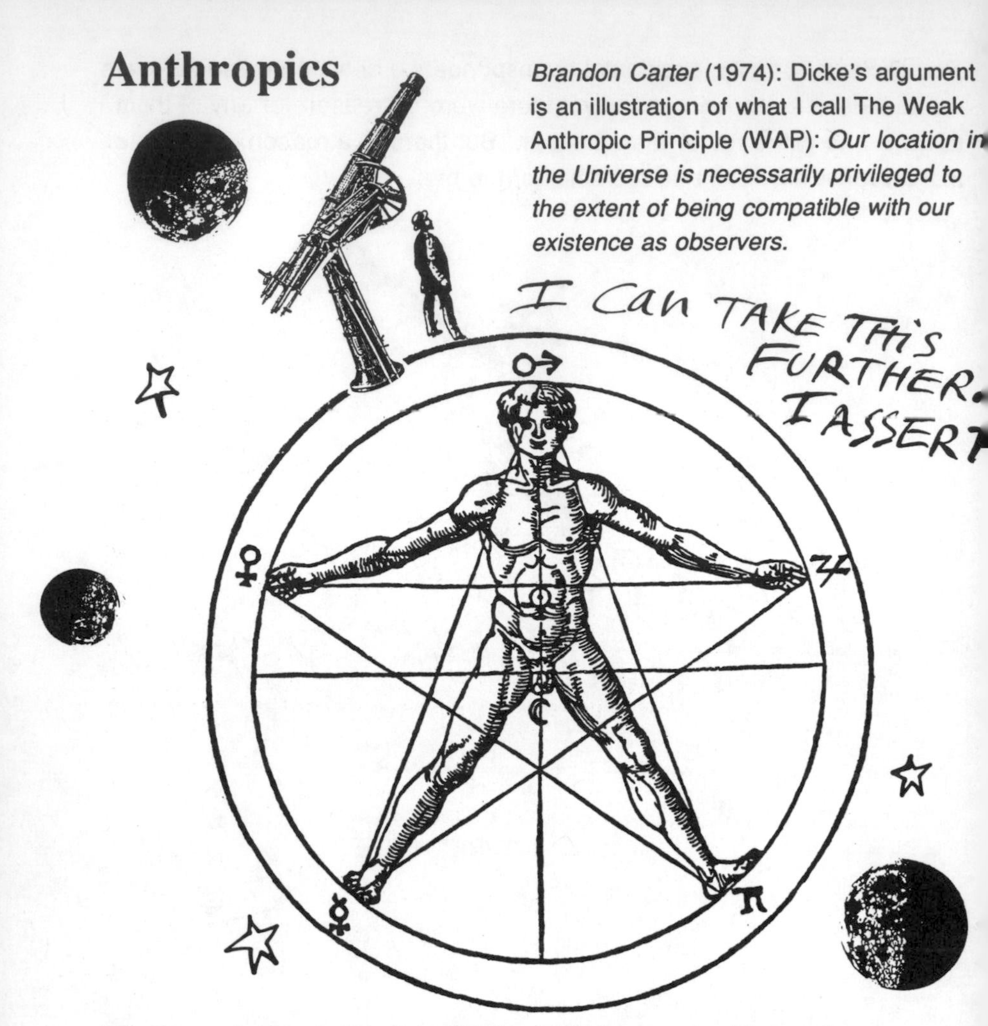


The Strong Anthropic Principle (SAP): *The Universe (and hence the fundamental parameters on which it depends) must be such as to admit the creation of observers within it at some stage.*

What could he have meant by "must"? He seems to have meant that from the existence of humans one could deduce restrictions on the constants of physics. But other people have taken it to mean that the Universe must have been designed so that people could evolve in it. Which leads them back to a belief in a Great Designer.

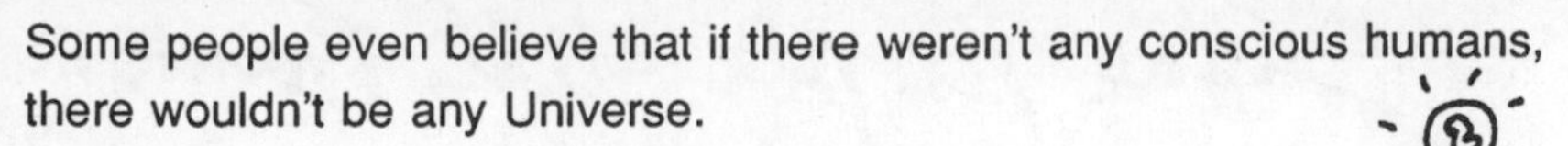

Some people even believe that if there weren't any conscious humans, there wouldn't be any Universe.

*Bishop Berkeley* (1710):

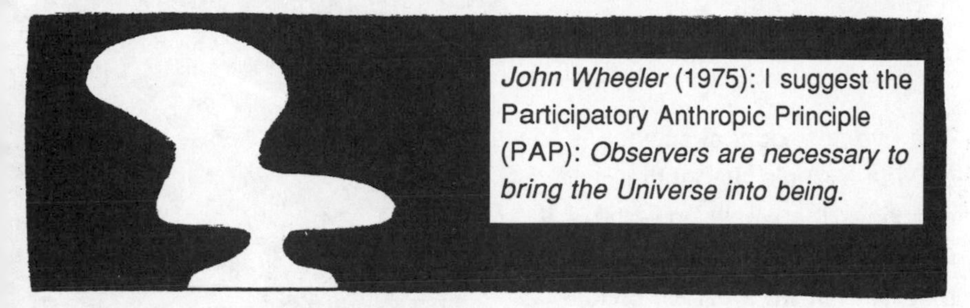


And some talk about the long-term consequences:

*John Barrow* and *Frank Tipler*: Here is the Final Anthropic Principle (FAP): *Intelligent information-processing must come into existence in the Universe, and, once it comes into existence, it will never die out.*

# Some people still Think...

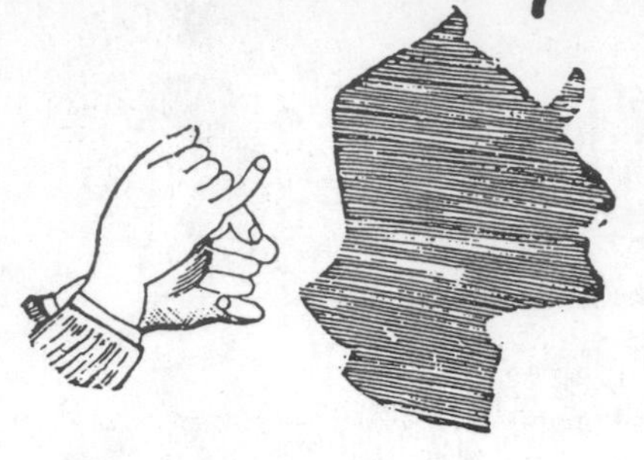

that the business of science is to find a Theory of Everything, answer "Why?" questions, and mix theology with cosmology.

The eventual goal of science is to provide a single theory that describes the whole universe.... if we do discover a complete theory, then we shall all, philosophers, scientists, and just ordinary people, be able to take part in the discussion of the question of why it is that we and the universe exist. If we find the answer to that, it would be the ultimate triumph of human reason — for then we would know the mind of God.

THIS 'ORDINARY PERSON ASKS: WHAT IF GOD HAS A CHANGE OF MIND!

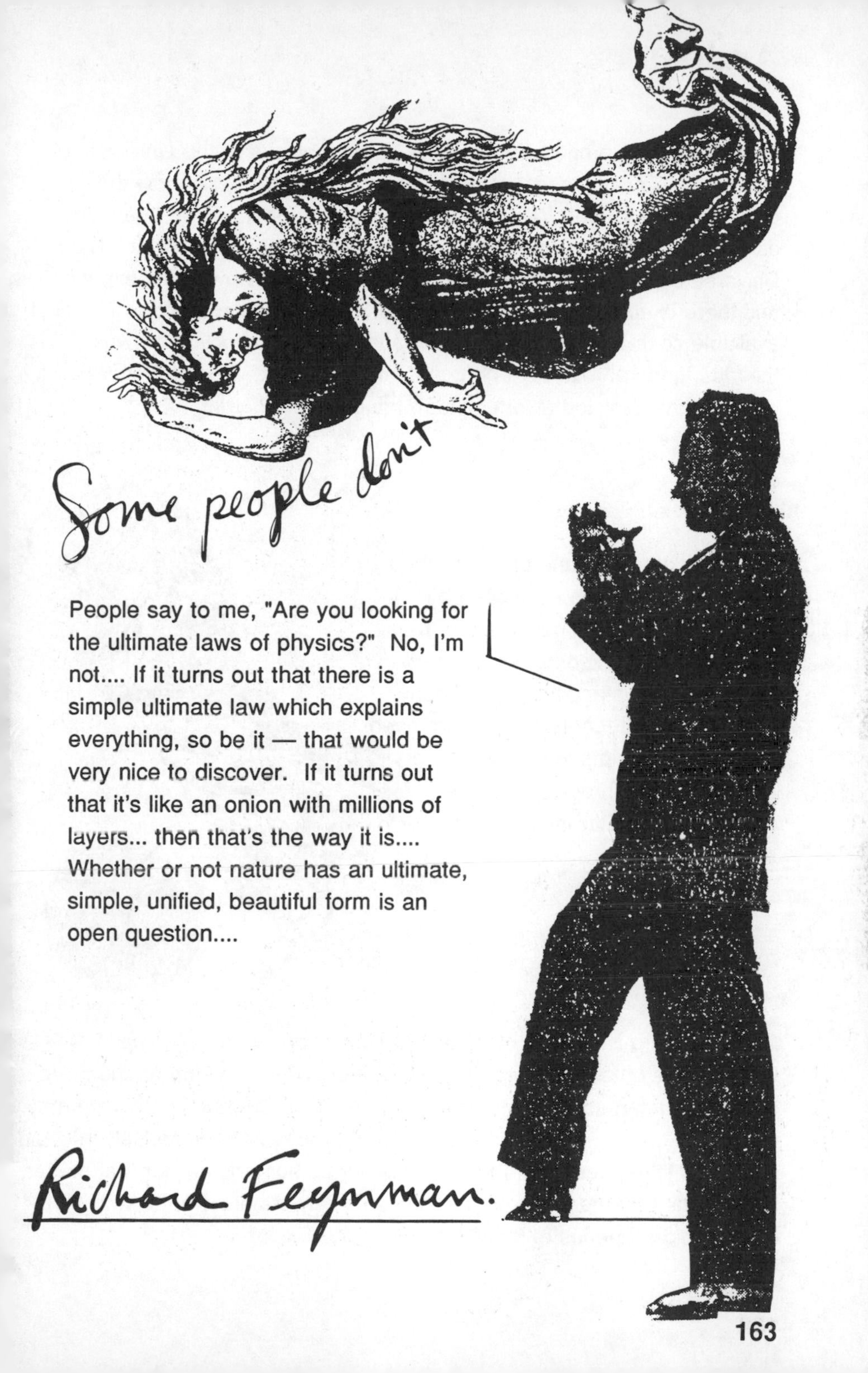
Some people don't
People say to me, "Are you looking for the ultimate laws of physics?" No, I'm not.... If it turns out that there is a simple ultimate law which explains everything, so be it — that would be very nice to discover. If it turns out that it's like an onion with millions of layers... then that's the way it is.... Whether or not nature has an ultimate, simple, unified, beautiful form is an open question....
Richard Feynman.

# And so...

Cosmology is in a critical state. The building blocks of the Universe are galaxies, but cosmologists have no clear understanding of how the galaxies were formed. To reconcile theory with observation, they must assume that galaxies are largely made of stuff they can't detect. The Universe expands, but they have no settled idea of how fast it expands, and there is no adequate theory of what started the expansion. The best available cosmological models are made from a concoction of two theories, general relativity and quantum mechanics, which for over half a century have resisted all attempts to bring them together.

In the Middle Ages and after, the theory of planetary motions developed by Ptolemy was in continual need of repair, until Copernicus and Kepler overturned it altogether and started afresh. Today, cosmological theory is in continual need of repair. Attempts to fix it, such as inflation and anthropics, have been unsatisfactory. Perhaps it too needs to be overturned.

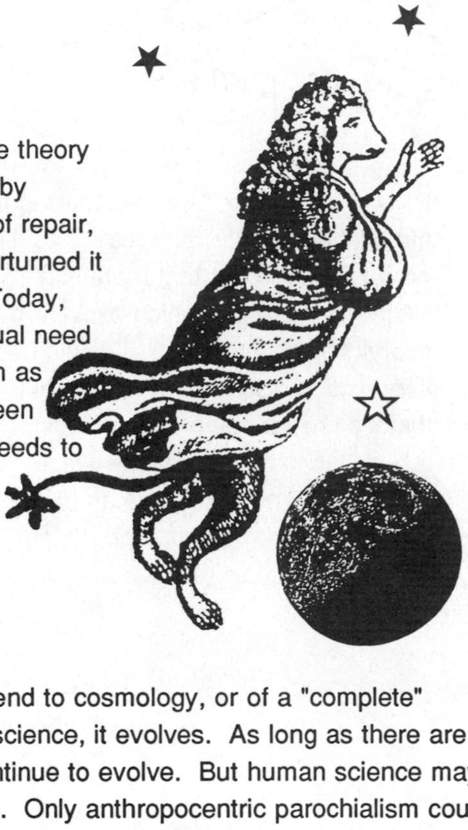

Certainly there is no sign of an end to cosmology, or of a "complete" theory. Like other branches of science, it evolves. As long as there are humans, human science will continue to evolve. But human science may not be the only possible science. Only anthropocentric parochialism could allow one to suppose that the Universe needs humans, or, for that matter, to presume that there are no non-human scientists elsewhere, with cosmological theories of their own.

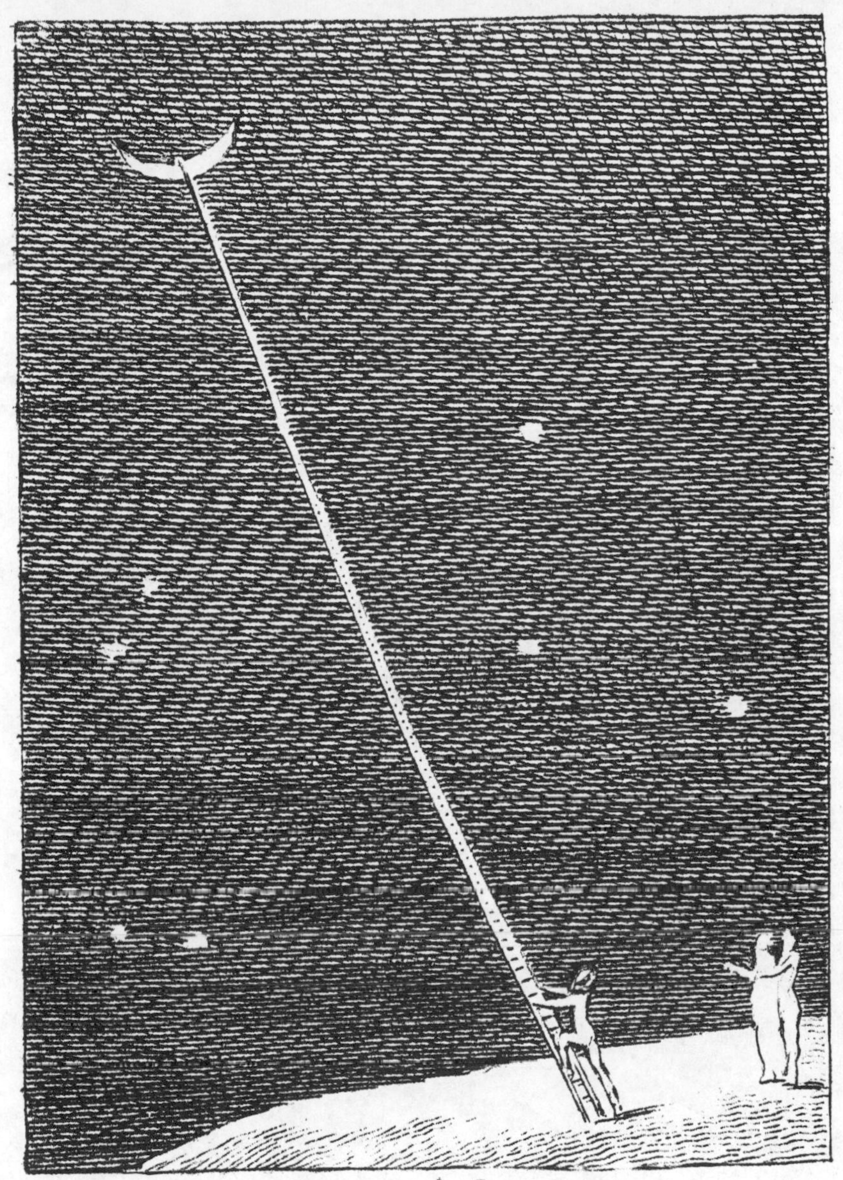


9 I want! I want!

We are the way we are because the Universe is the way it is ●●●

●●● and not vice versa

## Notes and References

In these notes, *The New Physics* = *The New Physics,* edited by Paul Davies, Cambridge University Press, 1989.

page

8 Richard Feynman's remarks are from his *Lectures on Physics* **1**, pp 1-2, 3-6.

13 A verse of "The Love Song of J. Alfred Prufrock" is quoted from T.S.Eliot's *Collected Poems 1909-1935*, Harcourt Brace, New York, 1936.

23 Vitruvius on the ants is adapted from M.H. Morgan's translation of Vitruvius, *The Ten Books on Architecture*, quoted by T.S.Kuhn, *The Copernican Revolution*, Vintage Books, 1959, pp 51-52.

25 C.S.Lewis *Miracles* (1947), Fontana Books edition, p 82. Harry Emerson Fosdick *The Modern Use of the Bible* (1924), pp 46-47.

26 Ko Hung is quoted from J. Needham, "The Cosmology of Early China", in *Ancient Cosmologies*, edited by Carmen Blacker and Michael Loewe, London, George Allen and Unwin, 1975.

27 The text, by Jinasena, is adapted from a translation of the *Mahapurana* (*The Great Legend*), in *Sources of Indian Tradition*, edited by W. Theodore du Bary, New York, Columbia University Press, 1966, **1**, pp 76-78.

29 The title of Copernicus's book translates to *On the Revolutions of the Heavenly Bodies*.

31 Luther and Calvin are quoted from A.D.White, *A History of the Warfare of Science with Theology in Christendom*, New York, Appleton, 1897, **1**, 126].

33 For Kepler's astrology see T.S.Kuhn, *The Copernican Revolution*, p 93.

36 For the prohibition of teaching about sunspots see A.D.White, *A History of the Warfare of Science with Theology in Christendom*, New York, Appleton, 1897.

36 For Father Clavius's theory of the moon see G. Geymonat, *Galileo Galilei*.

40 Galileo's abjuration is translated from *Le Opere di Galileo Galilei*, ed. A.Favaro, **19**, 406-407, 1909.

42 It is not true, as asserted by Paul Davies, *The Cosmic Blueprint*, Simon and Schuster, 1989, p 10, that Uranus was discovered by its gravitational effects on other planets.

50 Discovery of the bubble-like distributions of galaxies was reported by Valérie de Lapparent, Margaret Geller and John Huchra in *The Astrophysical Journal*, **302**, L1-L5, 1986. Discovery of the Great Wall was reported by Geller and Huchra in *Science*, **246**, 897, 1989.

54 Fred Hoyle's invention of the term "big bang", and his comment on it, are in his book *The Nature of the Universe*, Blackwell, Oxford, 1950, p 102.

55 "The name, Big Bang, is unfortunate..." is quoted from Peebles, Schramm, Turner and Kron, *Nature*, **352**, 769, 1991. For the limitations of current theory see "Quantum gravity", the chapter by Chris Isham in *The New Physics*.

64 The quotation about laws of nature is from Steven F Mason, *A History of the Sciences*, New revised edition, Collier Books, New York, 1962, p 173.

The comment by Eugene Wigner appears in his article "The Unreasonable Effectiveness of Mathematics in the Natural Sciences", *Communications on Pure and Applied Mathematics*, **13**, 1-14, 1960.

68 For more about quarks, see *The New Physics*, Chapter 14.

79 The phenomena of emission and absorption are more complex than Kirchhoff's

rules suggest. Many atomic processes may be active in a source of radiation.

82 For more on special relativity see *Einstein for Beginners.*

84 Strictly speaking, only a point has a world-line; an extended object is represented by a *world-tube* in space-time, consisting of the world-lines of all its points.

103 At velocities small compared to the velocity of light, the formula (for a uniform spherical object) is (escape velocity)$^2$ = $2GM/r$, where M is the mass of the object, $r$ its radius, and G the gravitational constant.

104 Among the candidates for black holes in this Galaxy are the object called Cygnus X-1 and two others, discovered in 1992, called V404 Cygni and Nova Muscae.

106 There will be earlier crises for the Earth, but it is likely to be habitable for at least another billion years. See K. Caldeira & J.F.Kasting, *Nature,* **360**, 721, 1992.

116 Chris Isham's comments are quoted from his chapter in *The New Physics.*

126 "Whatever started the expansion..." is from Peebles, Schramm, Turner and Kron, *Nature,* **352**, 769, 1991.

133 Freeman Dyson's speculations about the far future are published in *Reviews of Modern Physics,* **51**, 447-460, 1979.

134 Extracts from "Spain 1937" are quoted from Auden's *Collected Shorter Poems, 1930-1944* Faber and Faber, London, 1950. Auden omitted the poem from his *Collected Shorter Poems 1927-1957,* published in 1966. Richard Goodman's comment on it in the *Daily Worker* quoted from *W.H.Auden: the Critical Heritage,* edited by John Haffenden, Routledge 1983. For Auden's 1964 view see *Poetry of the Thirties,* collected by Robin Skelton, Penguin, 1964.

137 Engels's comment appears in a letter to H.Starkenburg dated 25 January 1894; see Dona Torr, ed., Marx and Engels *Selected Correspondence* London, Lawrence & Wishart, 1936, p 517.

141 Opinions about COBE are quoted from *Science,* **257**, 29, 3 July 1992 and The (London) Independent, 24 April 1992, p 1.

145 David Layzer quoted from his book *Cosmogenesis,* Oxford University Press, 1990.

146 For inflation, see the chapter by Guth and Paul Steinhardt, in *The New Physics.*

148 For dark matter see *The Fifth Essence,* by Lawrence Krauss, Vintage, 1990.

152 The hot-cold mix was proposed by Marc Davis, F.J Summers & David Schlegel, *Nature,* **359**, 393 and A.N.Taylor & M.Rowan-Robinson, *Nature,* **359**, 396, 1992.

156 For Eddington's first result see *Relativity Theory of Protons and Electrons,* Cambridge, 1936, for the second, *Fundamental Theory,* Cambridge, 1946.

158 Dirac's original paper is in *Proceedings of the Royal Society,* **A165**, (1938) 199, Dicke's in *Nature,* **192**, 440, 1961.

160 For Carter's definitions see his article in M.S.Longair, ed. *Confrontation of Cosmological Theories with Observational Data* Reidel, Dordrecht & Boston 1974.

161 Berkeley quoted from *Principles of Human Knowledge,* section 44. Wheeler, paraphrased by Barrow and Tipler, *Anthropic Cosmological Principle,* Oxford, 1986, p 22; Barrow and Tipler's FAP, *op. cit.* p 23. Martin Gardner, *New York Review of Books,* 1986, No 8 p 22.

162 Stephen Hawking quoted from *The Brief History of Time,* Bantam Press, London, 1988, pp 10 & 175.

163 Interviews with Richard Feynman quoted from *Genius: Richard Feynman and Modern Physics* by James Gleick, Little Brown, London, 1992, pp 432-433.

## Glossary

**absolute zero**: the lowest possible temperature, about —273 degrees Celsius; it can be approached, but not attained: 139

**acceleration**: rate of change of speed: 92, 94

**aether**: an invisible, weightless, crystalline solid which according to Aristotle fills the Universe, beyond the Earth: 20

**anthropic principle**: the idea that the place of humans in the Universe is not arbitrary, but must be consistent with the requirement that they should have had time to develop, and that the Universe must be such as to have allowed this: 160

**billion**: a thousand million; 10

**black body**: an object which can emit and absorb, but does not reflect, electromagnetic radiation; any distribution of matter in equilibrium with electromagnetic radiation, so that conditions are steady and as much is absorbed as is emitted, behaves as a black body: 110

**black hole**: an object so small for its mass that its gravitational field prevents the escape of matter or radiation: 103-104; believed to be the final state of evolution of a star of sufficiently great mass; quasars may be black holes: 110

**COBE**: COsmic Background Explorer satellite: 139-141

**comet**: a celestial object with a fuzzy head and a long tail which moves comparatively rapidly across the sky. Near the centre of the head there is a "nucleus" of starlike appearance. The nucleus is a small irregular solid object made of silicate dust, ice, and frozen organic compounds. Sunlight spreads some of the dust out to make the tail: 3,32

**deferent**: in Ptolemy's description of planetary motion: the large circle representing approximately the motion of a planet; the planet itself moves on an epicycle which is carried around the deferent: 24

**density**: the amount of mass in a unit of volume; the future evolution of the Universe is believed to depend on its average density: 115, 129, 144, 150

**deuterium**: "heavy hydrogen", an isotope of hydrogen with one proton and one neutron in each nucleus: 69, 127, 150

**electron**: a charged elementary particle; an atom consists of a nucleus surrounded by a cloud of electrons: 67-70

**elementary particle**: a basic constituent of matter: 53, 113, 146

**epicycle**: in Ptolemy's description of planetary motion: a small circle whose centre moves around the circumference of a larger circle; the larger circle may be another epicycle, or the deferent: 24

**galaxy**: a system of stars, gas and dust held together by gravitational forces: 45-50

**homogeneous**: having characteristics which do not vary from place to place: 122, 123, 125, 138

**infra-red**: electromagnetic radiation of frequencies just below those of visible light: 72

**isotope**: a particular kind of nucleus of an element with a particular number of neutrons: 69

**isotropic**: having the same appearance in all directions: 122, 123, 125, 138, 147

**Jupiter**: the largest planet in the Solar System: 23, 34, 35

**light-year** (ly): the distance light travels in a year, which is about 9.46 thousand billion ($9.46\times10^{12}$) kilometers: 7

**mass**: a measure of inertia, which means resistance to being accelerated; 92

**Megahertz** (MHz): a unit of frequency = one million Hertz, or one million cycles of oscillation per second: 75

**Mercury**: the innermost planet of the Solar System: 95

**nebula** (plural: nebulae, from the Latin word for "mist"): originally used to describe fuzzy or cloud-like celestial objects; those in which, with improved telescopes, stars have been distinguished, are now generally called "galaxies", while "gaseous nebulae" refers to clouds of interstellar gas and dust within this Galaxy: 42-45

**neutron**: an uncharged particle, one of the constituent particles of atomic nuclei, the other being the proton: 68

**neutron star**: a celestial object composed entirely of neutrons; believed to be the final state of evolution for a star of moderate mass: 108

**nova** (plural: novae, from the Latin word for "new"): a star which over a short period increases greatly in brightness: 32

**nucleon**: a neutron or proton: 69

**nucleus** (plural: nuclei, from the Latin word for "nut"): the central part of an atom, comprising nearly all of the mass: 67-70

**Planck density**: a density, about $10^{93}$ times the density of water, constructed from the fundamental constants of nature: 55, 115

**proton**: a charged particle, one of the constituent particles of atomic nuclei, the other being the neutron: 68

**quantum** (plural: quanta, from the Latin word for "how much"): a packet of energy: 110

**quantum mechanics**: theory of motion of material objects, especially relevant to systems of atomic size and smaller: 66, 113

**quasar**: quasi-stellar radio source: 110

**radiation, electromagnetic**: an oscillating electromagnetic field, propagated in space; forms of electromagnetic radiation include visible light, radio waves, infra-red and ultraviolet radiation, X-rays and gamma rays: 72-81

**radiation, gravitational**: an oscillating gravitational field, propagated in space: 100

**red giant**: a star larger and cooler than the Sun; the Sun is expected to evolve into a red giant: 106

**relativity theory**: (1) special relativity: theory of motion of material objects, especially relevant when there are relative motions at speeds comparable to the speed of light: 82-87; (2) general relativity: theory of motion of material objects, especially relevant on the scale of planetary systems and upwards and whenever gravitational fields are of prime importance: 66, 95-98

**solstice**: one of the two times of year when the Sun is highest or lowest in the sky: 16

**space-time**: in relativity theory, space (three-dimensional) and time (one-dimensional) considered together as the four-dimensional arena where physical processes take place: 82

**supernova** (plural: supernovae): a star which over a short period increases in brightness by a factor of a billion or more: 32, 107-110

**ultra-violet**: electromagnetic radiation of frequencies just above those of visible light: 72, 137

**white dwarf**: a small, dense hot star; the Sun is expected to evolve into a white dwarf after its red giant phase: 106

## Biographical Index

**Other books to read**

Besides the books referred to in the Notes, here are the names of a few others, out of the hundreds in the bookshops, which you may find interesting or useful:

Abell, George O., Morrison, David & Wolff, Sidney O. *Exploration of the Universe,* 6th edn., Saunders, 1991. Elementary college text, giving extensive background about stars and planets, as well as discussing cosmology.

Lightman, Alan & Brawer, Roberta *Origins. The Lives and Worlds of Modern Cosmologists*, Harvard University Press, 1990. Interviews with 27 cosmologists, including Geller, Guth, Hoyle and Hawking.

Pagels, Heinz R. *Perfect Symmetry*, Penguin 1992. Especially for the connections with elementary particle physics.

Riordan, Michael & Schramm, David *The Shadows of Creation*, Oxford University Press, 1993. Especially for dark matter and the connection with elementary particle physics.

Silk, Joseph *The Big Bang.* W.H.Freeman, 1988. Extensive discussion of the early Universe.

Weinberg, Steven *The First Three Minutes.* Flamingo Edition, 1983. Very clear introduction to the early history of the Universe.

The *Scientific American*, published monthly, frequently carries popular articles, without equations, about recent developments in cosmology and related subjects.

Felix Pirani is Emeritus Professor of Rational Mechanics in the University of London. He has published a number of scientific papers and popular articles about Einstein's theory of relativity, and three books for children. He was once attached to the group of cosmologists known as the Cambridge Circus.

Christine Roche is a French-Canadian cartoonist/illustrator who lives and works in London. She draws for various magazines and has illustrated several adult and children's books. Her books include *I'm not a Feminist, but...*, a book of cartoons, and *A Woman's History of Sex* (with Harriet Gilbert).

Felix Pirani and Christine Roche (neither of whom owns a dog) have collaborated on three other books. One of these was condemned by a Parliamentary motion for inciting children to alcoholism and violence.

**Acknowledgements**

Thanks to Richard Appignanesi, Monika Goppold, Bianca and Marta Monteleoni, J Ravetz, David Robinson and Chris Roper for their comments and to Frank Weil for his unstinting encouragement.

# Index